Spatial Optimization
for Managed Ecosystems

Complexity in Ecological Systems Series

Complexity in Ecological Systems Series

T. F. H. Allen and David W. Roberts, Editors
Robert V. O'Neill, Adviser

Robert Rosen
Life Itself: A Comprehensive Inquiry Into the Nature,
Origin, and Fabrication of Life

Timothy F. H. Allen and Thomas W. Hoekstra
Toward a Unified Ecology

Spatial Optimization
for Managed Ecosystems

John Hof and Michael Bevers

Columbia University Press
New York

Columbia University Press
Publishers Since 1893
New York   Chichester, West Sussex

Copyright © 1998 Columbia University Press
All rights reserved

Library of Congress Cataloging-in-Publication Data
Hof, John G.
　Spatial optimization for managed ecosystems / John Hof, Michael
Bevers.
　　　p.　cm. — (Complexity in ecological systems series)
　　Includes bibliographical references (p. 237) and index.
　　ISBN 0–231–10636–X (alk. paper)
　　ISBN 0–231–10637–8 (pbk.)
　　1. Spatial ecology—Mathematical models.　2. Ecosystem management—
Mathematical models.　I. Bevers, Michael.　II. Title.
III. Series.
　QH541.15.S62H64　1998
　577'.01'51—dc21　　　　　　　　　　　　　　　　　　　　97–25439

To Deb and Melanie

CONTENTS

SERIES EDITORS' FOREWORD

Timothy F. Allen and David W. Roberts

Robert Rosen's *Life Itself* and T. F. H. Allen and T. W. Hoekstra's *Toward a Unified Ecology* were the first two books in the series on complexity in ecological systems. Rosen's book laid out an abstract framework with great generality. Allen and Hoekstra's book was more overtly ecological, but shared Rosen's general model. Allen and Hoekstra's volume explicitly separated different types of ecology, each with its own chapter. They invited later volumes in the series to take up one or more of the chapter topics and make a book of them. There was also a significant emphasis on applied ecological issues in Allen and Hoekstra. The present volume works in applied ecology and treats spatial aspects of management planning. It is a book of applied landscape ecology addressing theory in management planning. It deals with the spatial aspects of topics as varied as pestilence, macrofauna movement, diversity, and water runoff, each of which occurs in a different type of ecological system, as defined by Allen and Hoekstra. Hof and Bevers cast populations, organisms, communities, and process-functional ecosystems in an explicit spatial framework. Their effort is to make predictions in systems that have a diverse set of types across a range of spatial scales. Hof and Bevers are relentless in pressing ecological problem solving in a complex setting as far as modern technology can take it. Their book is a splendid fit with the spirit of the series.

Hof and Bevers come from a school of landscape ecology in which the point of departure is mathematical programming models. They identify a set of constraints and solve the problem of meeting all those constraints simultaneously. The process is heavily computerized, the solution being found in an algorithmic fashion. Mathematical programming is a very important management tool, but not widely known across ecology at large. Coming from a background in economics and management theory, these authors bring a distinctive point of view to spatial ecology.

Hof and Bevers may well not view themselves as landscape ecologists at all, but they are. The literature that they cite is closely trained on the problems they address and the general approach their methods prescribe. There are central ideas in ecology and cultures of data analysis that Hof and Bevers do not address, or only give a nod. This gives their book a fresh approach, unhampered by any need to weave into an intellectual tradition that is not their own. Accordingly, their book is distinct from, but runs parallel to, modern landscape ecology as it is practiced in North America.

Hof and Bevers do cite familiar names in ecology, such as Diamond, Karieva, and Simberloff, who are broad in their outlooks but who share a certain mechanistic style of modeling that assumes a population ecological posture. As a result, scholarly as this volume is, its connections to the larger body of landscape ecology are intermittent. The editors of this series have therefore taken up the authors on their invitation to throw a line across to the rest of ecology, which may be more familiar to a segment of the intended readership. We do not have to throw the line very far because the course of this volume is close to many aspects of ecology, landscape ecology in particular. The general approach of Hof and Bevers will be very helpful to many spatially oriented ecologists, but it may be new enough to miss some of that audience. That would be unfortunate, for the learning curve is not that steep. The chapters are modular, and iterate around the same sequence of subheadings, but each chapter has a new topic and class of problem. Get through the first chapter, and you have a model for mastering the other techniques Hof and Bevers have developed. It is our job in this foreword to flag some connections to the familiar, so that it all seems less alien. Soon enough, Hof and Bevers will get landscape ecologist readers in tow, for they show us how to solve diverse problems.

Spatial arrangement of ecological material has been a central concern, even before ecology was a self-conscious discipline. The biogeographers of the last century, such as von Humboldt, were looking at spatial systems on a very large scale. At the turn of the century, American ecologists Clements and Cowles were forging a new discipline of community ecology, using a spatial reference as their point of departure. They both introduced vegetation dynamics, Cowles (1899) on the sand dunes by

Lake Michigan and Clements (Pound and Clements, 1900) in the vegetation of Nebraska. Both works referred to spatial arrangement in vegetation, so that the dynamics had a context. Clements, in particular, developed the quantitative use of quadrats spaced across the study site. Alas, no computation was available when the seminal works were written. There is plenty of computational power now. The approach of Hof and Bevers can be seen as a quantification of the dynamics between quadrats in a fashion that Clements himself might have wished he could do. In Britain, Tansley (1939), reporting work of Pearsall, was able to equate spatial arrangement of vegetation with waves of vegetation maturation, in what are called vegetation seres. These patterns would be very amenable to the analysis of change over time across space, which is a significant part of the present volume.

After the early efforts to interpret vegetation in a spatial framework, remarkably few papers turned on spatial issues. However, there were notable exceptions. The tour de force midcentury was Alex Watt's (1947) paper. Watt pulled together about eight examples of vegetation in cyclical patterns of development and decline. For each example a particular spatial pattern emerged. Variants of these patterns were cited from different locations across a range of vegetation types: bog hummocks, Calluna bands on hillsides, grassland hummock building, and forest gap phases. Hof and Bevers have an explicit section on autocorrelation, which could address vegetational patterns of the class that Watt considered. Peter Greig-Smith (1952) pioneered autocorrelation techniques shortly after Watt's seminal work, and methods have been further elaborated since then to include Fourier analysis and even wavelets.

It is unfair to say that landscape ecological approaches disappeared through the middle decades, for although plant ecologists generally moved away from explicitly spatial organizing principles, wildlife ecologists maintained a limited interest in spatial issues. The black-footed ferret example in this text follows this tradition, but presses the issue of prediction much harder.

Spatial research papers, after the early classics, are scarce because either the pattern is clear or it overwhelms unaided human capacity for memory or data analysis. The early classic papers in spatial ecology did what computationally unaided ecologists could do. But after that, the leaders in the field went off to do other types of ecological investigations using community and population criteria. The modern revival in spatial ecology had to wait until the advent of modern computation. Complex patterns are now commonly treated quantitatively. The present volume is part of the revolution in information processing. One could not begin to solve problems in the manner of Hof and Bevers without massive computer power.

A great strength of the present volume is its introduction segments for new topics. References to the mainstream of modern American landscape ecology or to the ecological intellectual culture of multivariate description of vegetation are few. Even so, the main topics covered in this work are quite closely related to the mainstream of quantitative landscape ecology. Modern landscape ecology in North America is distinct from the tradition of vegetation mapping of heavily manicured landscapes in Europe. The distinction is the muscular quantification on the North American side. The tradition in mathematical programming is to make explicit assumptions in any given analysis: "If you want to make such and such assumptions, then the optimum solution between competing constraints is this." By contrast, the quantification in the vanguard of North American landscape ecology is more open-ended. That is not to say that the papers in landscape ecology lack an explicit statement of assumptions, but it is often achieved with what are called neutral models of a general sort, for the purposes of comparison to some observed situation.

Neutral models are the embodiment of the assumptions in a given study. The neutral models used in wildlife ecology midcentury, such as the random walk, were simpler than their modern counterparts, but in the same spirit. When the result of a set of field observations is interpreted with the random walk, the default setting (the "as opposed to what" factor) is random movement anywhere on the landscape. Compare this with the neutral models of Bob Gardner and his colleagues, coming from percolation theory (Gardner et al., 1987, 1989). The application of percolation theory comes from soil science and material science. The percolation is of water through air spaces in soil, or of electricity in an amalgam, where only one component conducts. When the proportion of the whole that allows passage reaches a critical threshold, the water or electricity passes through the whole for an indefinite distance. In percolation models, the comparison of field data is to a landscape where there is random insertion of patches of landscape that allow passage of an animal or a disturbance. The randomness in a random walk allows the animal to go anywhere. The percolation model is still random, by virtue of the random placement of percolation pixels, but it ties the animal to variation in patches within the landscape. Thus the percolation neutral model is centered on landscape properties, not on movement of the animal or the agent of disturbance. The random walk is subsumed in the percolation model because the landscape agent moves so thoroughly at random that, if there is a way across, it will find it. The assumptions of the various models in Hof and Bevers are, at different times, similar to both the percolation and the random walk models. The real distinction is that the percolation and random walk models typically take a habitat layout as a given, whereas Hof and Bevers treat the layout as a choice variable.

Close as the Hof and Bevers models are to percolation and random walk models, there are differences in style. Bruce Milne drew attention to the gap (Milne, 1987) in his comparison between percolation models and the original Forest Service optimizing model, FORPLAN. Note the generality of the early percolation model of Bob Gardner and his colleagues in landscape ecology. With the Gardner model one can say that material landscapes percolate at a lower threshold of occupiable habitat than neutral landscapes. The models found in this volume are used in a less general and more focused manner. Thus the modeling style of Hof and Bevers gives more insight into the particulars of almost any specific situation one could imagine. Later percolation models of Milne and colleagues do move in the direction of particularity that is manifested in this volume (Milne et al., 1996). If Milne changes the rules that connect adjacent occupiable habitat, he can move the percolation threshold to any specific isopleth of pixel density on the landscape. Milne's particular rules of connection do the same job as the wonderfully explicit assumptions of Hof and Bevers. One cannot have it both ways, and this volume puts more emphasis on being explicit and particular than on making statements that apply generally to landscapes in principle. It is telling that modern percolation models are moving in the same direction, although there is still much greater particularity in the assumptions of Hof and Bevers.

All science is about the investigation of assumptions. One is less interested in the verity of assumptions because all assumptions are arguably false. One learns more from the failure of assumptions than from being able to get away with them. The failure takes two forms, and both are informative. The difference turns on whether the assumptions are reasonable. One learns when unreasonable assumptions fail to give unreasonable results, or when reasonable assumptions fail to give reasonable results. Let us see where Hof and Bevers fit in this scheme of things, and who are their close neighbors.

When one makes patently false assumptions, informative failure occurs as the model fails to crash, when one might expect it to do so. This is a favorite style of modeling in population ecology. Compare the styles of Monica Turner and Tony Ives, both colleagues of Allen at the University of Wisconsin. Turner is a former president of the North American chapter of the International Association for Landscape Ecology, and Ives is a population biologist. When Turner gave a talk at the University of Wisconsin on her work on ungulates in Yellowstone, she made reasonable assumptions about ungulate movement. In the audience was Ives, who said later that he would have had those deer moving in the model at the speed of light. His strategy would have been to find out whether movement mattered much at all. Ives's deer would require a correction for relativity. The failure of such creatures to give unreasonable results would indicate that movement of deer is an unnecessary parameter, a very useful thing to know if it were true. Turner's approach to challenging assumptions is entirely different, and represents the school of landscape ecology that uses rich simulations. She makes reasonable assumptions and learns when, despite her best efforts, the model fails to be close to field observation. She then learns that the reasoning behind her reasonable assumptions is incomplete or somehow flawed. In their style of testing assumptions, Hof and Bevers are much more allied to Turner and other landscape ecologists. But there are differences, for Hof and Bevers's quest is to find a very explicit set of assumptions. The power of mathematical programming is that elaborate assumptions are laid out in exquisite detail. Then you can work out whether you want to make those assumptions.

In the social sciences there is a polarity between economists and sociologists. Economists create a model, and only then go out to test the model against material economic happenings. Economists may not perform the test on any material system at all. By contrast, sociologists usually start with only a general arena of discourse, and immediately start to collect data. Their genius is in analyzing data skillfully to see patterns, which will invite a specific model. As one might expect, Hof and Bevers write and theorize more like economists than sociologists. In ecology, there is a tradition that is so like sociology that is calls itself phytosociology. The modern summary work in the field is Legendre and Legendre (1987), and the methods are called ordination and classification. The iconic names in the field include John Curtis, David Goodall, and Robert Whittaker. Just as sociologists can profitably team with economists, the tradition of phytosociology in ecology could benefit from teaming with Hof and Bevers.

Phytosociology receives less attention now, such that the prevailing neglect in system description leaves many ecological systems inadequately described before the experimentation starts. This could also be the case for optimizing ecological systems, which could be viewed as another means of ecological experimentation. Even with modern computers, we are close to the limits, despite using austere characterizations of ecological systems. The tradition of data reduction in phytosociological vegetation studies should prove to be a very useful means of reducing the number of parameters in the system, thus keeping optimizing techniques within computational limits. Conversely, the optimizing methods here may be a very helpful adjunct that allows vegetation description to move into a dynamic landscape mode.

The final area of discourse related to the methods and problems laid out in this volume comes from simulation of vegetation. There are many forest stand simulation programs, derived from Hank Shugart's FORET stand simulator (Shugart, 1984). The strategy is to grow individual plants without any spatial assignment, and submit each one to a probability of dying, surviving, or growing, depending on environmental conditions. The probabilities are assigned to each species separately. Stand simulators have been remarkably successful in generating forest data that appear realistic. They are portable to forests even on other continents, by reassigning growth and shade parameters to species that appear on the new continent. FORET grows a given area of forest, not by having space assigned explicitly by area, but by having limits on standing crop. There have been elaborations that include soil minerals, beyond the moisture and temperature factors that occur in the first generation models. Of course, none of this would apply to spatial optimization, were it not for a metamodel version that grows stands in adjacent patches, allowing a patch to influence its neighbors. ZELIG is a metamodel of that sort developed by Smith and Urban (1988) and discussed in Urban et al. (1991). ZELIG is the computer-generated equivalent of Watt's (1947) gap phase vegetation. With increases in computational power, SORTIE belongs to a new class of individual-based models for forests that can feasibly include spatial position of each individual (Pacala et al., 1993). Recent work with SORTIE predicting the spatial recruitment of seedlings (Ribbens et al., 1994) appears to invite comparison with the optimizing models of Hof and Bevers.

In this foreword, we have drawn connections across to various classes of spatial models in ecology. It appears that the spatial questions posed in this book are clearly related to various spatial modeling issues across ecology. The optimization models herein often share classes of problems across mainstream landscape ecology. Despite its pertinence to a wide range of spatial considerations, this book is very coherent and focused. It offers a specific methodology and a class of methods that generally do not receive the attention they deserve in spatial ecology, outside the management literature. Mathematical programming and optimization within a set of chosen constraints is something most spatial ecologists could use. This book makes a compelling case for the use of these methods. Nowhere else in spatial ecology is there a general protocol for making such elaborate and explicit choices about the system under investigation. This book should have a wide audience across landscape ecology and in the spatial aspects of ecology at large.

References

Cowles, H. C. 1899. The ecological relations of the vegetation of the sand dunes of Lake Michigan. *Botanical Gazette* 27:95–117, 167–202, 281–308, 361–391.

Gardner R. H., B. T. Milne, M. G. Turner, and R. V. O'Neill. 1987. Neutral models for the analysis of broad-scale landscape pattern. *Landscape Ecology* 1:19–28.

Gardner R. H., R. V. O'Neill, M. G. Turner, and V. H. Dale. 1989. Quantifying scale-dependent effects of animal movement with simple percolation models. *Landscape Ecology* 3:217–227.

Greig-Smith, P. 1952. The use of random and contiguous quadrats in the study of the structure of plant communities. *Annals of Botany* 16:293–322.

Legendre, P., and L. Legendre, eds. 1987. Developments in numerical ecology. *Proceedings of the NATO Advanced Research Workshop on Numerical Ecology*. NATO ASI Series G: *Ecological sciences,* vol. 14. New York: Springer Verlag.

Milne, B. T. 1987. Hierarchical landscape structure and the Forest Planning Model. In *FORPLAN: an evaluation of a forest planning tool.* USDA Forest Service General Technical Report RM-140, pp. 128–132.

Milne, B. T., A. R. Johnson, and T. H. Keitt. 1996. Detection of critical densities associated with Pinon–Juniper woodland ecotones. *Ecology* 77:805–821.

Pacala, S., C. Canham, and J. Silander, Jr. 1993. Forest models defined by field measurement I. The design of a northeastern forest simulator. *Canadian Journal of Forest Research* 23:1980–1988.

Pound, R., and F. Clements. 1900. The phytogeography of Nebraska, 2nd ed. Lincoln, NE: The Seminar.

Ribbens, E., J. Silander, and S. Pacala. 1994. Seedling recruitment in forests: calibrating models to predict tree seedling dispersion. *Ecology* 75:1794–1806.

Shugart, H. H. 1984. A theory of forest dynamics. Springer Verlag, New York.

Smith, T. M., and D. L. Urban. 1988. Scale and resolution of forest structural pattern. *Vegetatio* 74:143–150.

Tansley, A. G. 1939. *The British Isles and their vegetation.* Cambridge University Press, Cambridge, England.

Urban, D., G. Bonan, T. Smith, and H. H. Shugart. 1991. Spatial application of gap models. *Forest Ecology and Management* 42:95–110.

Watt, A. S. 1947. Pattern and process in the plant community. *Journal of Ecology* 31:1–22.

PREFACE

This book presents ideas and methods for directly optimizing the spatial layout of management actions across the landscape within which an ecosystem functions. Capturing spatial considerations is motivated by the fact that a great part of an ecosystem's structure and function is spatial in nature. In turn, the emphasis on the ecosystem is motivated by the evolution of natural resource management away from a traditional agricultural problem to one where increased importance is placed on the health of the ecosystem itself.

The problem is complex, and the book relies heavily on mathematical presentations. However, the reader can skip over the math and still get the general idea. Examples are provided in all chapters. The book is written for advanced undergraduate and graduate students and is also intended for scientists in ecological, economic, and operations research fields, practicing natural resource planners, managers, and modelers. We hope that ecologists will forgive our simplifying assumptions and give the book some attention, because so little optimization work has been done in the ecological sciences. Familiarity with at least the basics in optimization (mathematical programming) as historically applied to natural resources is assumed. Treatments of these basic applications abound in the literature, so the reader can easily obtain more background along these lines.

The book is organized into four parts. The first treats simple, static, spatial relationships. The second treats stochastic spatial relationships as they manifest themselves in spatial autocorrelation. The third treats spatial changes in ecosystem components over time. The fourth discusses diversity and sustainability considerations. Throughout the book, spatial wildlife habitat considerations tend to dominate, but we also include chapters on a recreation problem, a water runoff problem, a pest management problem, and several aspects of timber management. The book is not a comprehensive treatment of the subject; rather, it is an introduction.

This book is put together largely from material we have previously published in smaller pieces. To our coauthors of these previous works (identified with a footnote at the beginning of each chapter), we are gratefully indebted. We also thank the USDA Forest Service, especially Fred Kaiser, Denver Burns, Tom Hoekstra, Brian Kent, and Marcia Patton-Mallory, for supporting this work. Many thanks to Dorothy L. Martinez and Lynn Meisinger, who worked so hard to complete the word processing through many revisions, to Joyce Patterson for illustrations, and to Jill Heiner for computer support. Finally, we wish to thank Timothy Allen and Robert O'Neill for their very helpful reviews of the manuscript.

Spatial Optimization
for Managed Ecosystems

Chapter 1

INTRODUCTION

Viewpoint

The ecological sciences are placing increasing emphasis on the importance of spatial factors in understanding forest and other wildland ecosystems. For example, the Special Feature of Vol. 75 (1) of *Ecology* is titled "Space: The Final Frontier of Ecological Theory." At the same time, public policy regarding the management of wildlands has begun to recognize that the functioning of the ecosystem must be taken into account and given high priority. A great deal of work is developing in the simulation literature that focuses on spatial considerations, but the topic has barely been touched in the optimization literature. Certainly, the choice between simulation and optimization methods should be based on the problem at hand. Simulation approaches are almost always able to handle more detailed representations of ecological functions, but optimization approaches can implicitly evaluate huge numbers of options and allow tradeoff analyses that might otherwise be impossible. For more reading on the use of optimization in multiple-resource management, see Hof (1993). This book examines the use of optimization in the management of an ecosystem, with the objective of directly capturing spatial ecosystem relationships and processes. This is a new field, and the book is exploratory in nature. Mathematics is used extensively but is balanced with simple examples.

In the forestry literature, the predominant perception of spatial considerations is that of adjacency relationships. Concerns regarding adjacent timber harvests emerged from legal and regulatory restrictions on the effective size of harvests; for example, if two seemingly legal-size clearcuts occur within a short time of each other in adjacent areas, the effective size of the clearcut may not be legal. Because the original motivations for the size limits included concerns for aesthetics, wildlife habitat, water quality, and so forth, addressing adjacency restrictions has often been interpreted (or even advertised) as addressing those underlying concerns. A fundamental thesis of this book is that addressing the actual spatial issues requires capturing the specific spatial relationships or processes involved, and that adjacency constraints are adequate only when the problem is undesired adjacency of management actions per se. Avoiding adjacency of management actions may actually be counterproductive for connecting wildlife habitat patches, creating edge, controlling the spread of exotic pests, managing water flows, or managing spatially defined uncertainty. The best way to address such concerns is to capture them directly in the analysis. In attempting to do so in this book, we often must make specific (sometimes simplistic) assumptions about how ecosystems function, but we believe that this is a more productive path than very precisely meeting criteria that address the actual problem indirectly and sometimes inconsistently. For those who face adjacency restrictions as the actual problem, the topic is dealt with extensively elsewhere and receives no further attention here.

Another traditional forestry problem that might be viewed as being spatial is that of roading networks and collection systems in harvesting operations. Because the topic of this book is capturing the spatial relationships and processes of the managed ecosystem, this subject is also left to other sources. Like adjacency restrictions, roading and collection systems for timber harvesting have been treated extensively elsewhere.

Organization

Parts I–III address the three categories of spatial relationships and processes that we can identify at this time. The introduction to each part describes the basic concepts and problems in general terms; the subsequent chapters use specific examples to demonstrate the various methods in optimization models.

Part I treats static spatial relationships. The introduction describes these as the basic relationships that reflect the importance of how close things are to each other, what shape things are, the difference between many small things and few large things, and so on, all interacting in a system across the landscape. Chapters 2 and 3 discuss wildlife habitat examples and treat edge effects, degrees of connectivity, and size thresholds in wildlife habitat patches that support a metapopulation. Chapters 2 and 3 use cellular (raster) and geometric (vector) management variables, respectively, to capture these relationships in an optimization model. Chapter 4 discusses a very different type of static spatial relationship—an economic one—with a recreation example. This model optimally locates recreation sites so as to find a spatial equilibrium between supply and demand. Recreation supply is spatial because potential sites are specifically located where resources support different activities. Recreation demand is spatial because potential recreators are concentrated in population centers. The model accounts for the spatial nature of supply and demand in strategically locating recreation sites in an efficient supply–demand equilibrium.

Part II treats the phenomenon of spatial autocorrelation in a chance-constrained modeling framework. The introduction defines the problem and chance-constrained approaches to it. Chapters 5 and 6 are distinguished, as chapters 2 and 3 are, by the cellular and geometric management variables. In this case, chapter 5 is not specific to a wildlife habitat example, but chapter 6 is, so that the spatial autocorrelation treatment can be combined with habitat connectivity (and the tradeoff examined). The methods used in chapters 5 and 6 are necessarily nonlinear, so chapter 7 is included to suggest approaches when nonlinear methods are not practical.

Part III treats the spatial processes that take place in forest ecosystems, primarily movement over time. The introduction clarifies the relationship between ecological system simulation models and the optimization approaches discussed here. Chapter 8 captures wildlife population growth and dispersal over time in a cellular optimization model, with multiple habitat seral stages (or types) that are affected by management actions, but also change over time on their own. In this problem, either integer or continuous management variables are often usable and the implications of each are explored. Chapter 9 provides a real-world example for the special case of the black-footed ferret releases in South Dakota, where management actions involve creating (by not poisoning) prairie dog towns for ferret habitat. The optimal placement of future ferret releases is also explored. Though much larger than the model discussed in chapter 8, this model is simpler in that only one habitat type is required, and it does not change with time by moving through seral stages. This case study was developed to support the federal land management plan revision for the Nebraska National Forest and to assist with black-footed ferret reintroduction in Badlands National Park over the next several years. The study was performed cooperatively by the Rocky Mountain

Forest and Range Experiment Station and the Nebraska National Forest of the USDA Forest Service. Chapter 10 describes a model that optimizes the placement of pest-control management actions over time, accounting for the pest population's growth and dispersal processes. This model is quite different from the modeling approaches in chapters 8 and 9 because it attempts to *reduce* population growth and dispersal. Chapter 11 addresses the spatial aspect of water flow as it relates to vegetative manipulation. In contrast with the water models that focus on localized impacts on stream segments (or reaches), this chapter uses a nested schedule in which vegetative manipulation is scheduled over a long time frame and water flow in a targeted storm event is modeled over a much shorter time frame. The spatial arrangement of different vegetative states (that result from the vegetative manipulation) affects the rates of water flow through the watershed in the storm event; thus the confluence of damaging volumes of water at various locations can be mitigated. All the models in part III are different, but all track movements across the landscape over time. Because of the dynamic complexity, all the models use the cellular management variable definitions.

In parts I–III, we focus on capturing spatial aspects of ecosystem processes, without much regard to our management objective. Current objectives in ecosystem management tend to emphasize considerations such as biological diversity and sustainability. Part IV focuses on capturing these considerations in objective functions, as important parts of managing an ecosystem. The introduction identifies faunal species richness as the most tractable measure of ecological diversity for forming optimization objective functions. Alternatives such as genetic diversity and ecosystem diversity are briefly discussed. Chapter 12 develops specific species-richness objective functions along with piecewise approximation procedures for use in linear models. These functions are then incorporated into the spatial model described in chapter 8. Chapter 13 provides an analysis of optimized species-richness equilibria (point and cyclic) that are sustainable over the long term. This analysis requires the inclusion of many species and many time periods (with a long time horizon), which in turn necessitates de-emphasis of the spatial considerations. In this sense, this chapter is a departure from the rest of the book, but we cannot currently handle spatial detail, many species, and long time frames all in one model.

Our inability to collect all the considerations discussed in parts I–IV into one model clarifies the current state of the art in this new area of research. Chapter 14 attempts to synthesize the concepts and methodologies described in the book, in this context. This chapter also discusses optimization as part of an adaptive management process that reflects our limitations in understanding, analyzing, modeling, and managing ecosystems.

Methods

At this time, we have explored two ways of depicting spatial options. One is with a cellular grid. The cell size with this option must be fine enough to capture the spatial concerns because choice variables are defined for each of the cells and irregular shapes are approximated with aggregations of the cells. The second way is with geometric shapes. Here, the choice variables define the size and location of the geometric patches. The basic shape of each patch must be prespecified, but includes shapes such as rectangles, allowing the height and width to be choice variables as well. Irregular shapes can be approximated with sets of these prespecified shapes. For example, an L-shaped patch can be approximated with an appropriate row and column of circles or with two appropriately sized and placed rectangles. Each of the two ways of depicting spatial options has its advantages and disadvantages, laid out in the chapters that follow. At the outset, the most important point is that the models attempt to optimize the spatial layout of management actions and state variables, directly including the spatial relationships and processes that are important in the managed ecosystem.

Some of these problems are solvable with linear methods, but many require integer or nonlinear methods. In some cases, integer variables can be used to capture nonlinear interactions quite effectively. For example, suppose two continuous variables, x and y, interact such that $x = \sum_i b_i z_i$ when $y < p$ (where p is a constant) but $x = 0$ when $y \geq p$. We can capture this with an integer $\{0, 1\}$ accessory variable, q, and an arbitrarily large constant, M,

$$x + Mq \geq \sum_i b_i z_i, \tag{1.1}$$

$$q \leq y/p,$$

$$x \geq 0,$$

as long as the general formulation tends to minimize x.[1] This type of approach is very useful for threshold effects or choice variable interactions that cannot be captured with simpler limiting factor constraint sets.

---

1. Throughout the book, variables should be assumed to be restricted to nonnegative real number values except when otherwise explicitly defined.

Solvability of integer and nonlinear programs is discussed briefly below, but it should be pointed out that it is the *combination* of nonlinear functions and integer variables that is particularly difficult to solve. Therefore, we present no formulations with this combination (although it is tempting to do so at times). When one already has integer variables, there are often ways of avoiding nonlinearities. For example, an interactive nonlinear constraint such as

$$xy \leq 0, \tag{1.2}$$

$$x \in \{0, 1\},$$

$$y \in \{0, 1\},$$

which specifies that either x or y (or both) must be 0, can be modeled with integer variables in a linear constraint as

$$x + y \leq 1. \tag{1.3}$$

Or, if one needs the product of x and $y \in \{0, 1\}$ it can be obtained in variable s by

$$s \leq x,$$
$$s \leq y, \tag{1.4}$$

as long as the overall formulation tends to maximize s. We use these types of formulations in the book to avoid the combination of nonlinear functions and integer variables, but these logic constraints often make the integer models more difficult to solve, as discussed next.

Solvability of (0–1) Integer Programs

Integer programming is similar to linear programming except that at least some of the variables are constrained to be integers. In this book, we use only 0–1 integer variables, which must assume a value of either 0 or 1. Most operations research texts treat integer variables, and Plane and McMillan (1971) devote an entire book to the subject. The examples provided here were all solved with readily available commercial solvers. Many commercial linear programming solvers have integer variable options.

The solvability of any given integer program is difficult to predict. The number of integer variables is certainly a factor, but the overriding determinant is how "intrinsically integer" the problem solution is. Most commercial solvers (which typically use a branch-and-bound algorithm) start with a linear programming solution that replaces the 0–1 restriction with a lower bound of 0, an upper bound of 1, and a continuous variable designation. The proximity of this first solution to an integer one is probably the best predictor of integer programming solvability. It is also important to note that this initial linear programming solution and similar ones that recur during the solution process provide bounds on the integer solution (because adding the integer restrictions can never improve objective function attainment). Thus, even if an optimal solution cannot be found, a satisficing solution within some tolerance of suboptimality is often achieved. This is quite useful if the bound is close to whatever integer solution can be found, which typically happens when the linear programming solution is nearly integer. The reader is warned that many of the instrumental variables and logic constraints used in this book tend to make the program less intrinsically integer and can render the bounds on suboptimality useless. As solution procedures continue to improve (along with raw hardware power), these problems can be expected to diminish. We can look back to when linear programs of modest size were very difficult to solve on the best mainframe computers, and feel more confident that the time will come when integer programs of considerable size and complexity will be solved as routinely as very large linear programs are now solved on personal computers.

Solvability of Nonlinear Programs

Nonlinear programming is similar to linear programming, except that nonlinear functional forms are allowed in the objective function and in the constraints. A number of good textbook treatments of nonlinear programming are available (such as Luenberger, 1984). An excellent treatment in the economic context is in Chiang (1984). The examples provided were solved using the generalized reduced gradient (GRG2) algorithm (Abadie, 1978). This is a very powerful search algorithm that finds, like all nonlinear programming solvers, a *local* optimum.

To get an intuitive feel for local versus global optima, see figure 1.1. Assume that we wish to select a value for a choice variable, x, so as to maximize an objective function, $f(x)$. Obviously, the true or global optimum is at point b. However, a nonlinear programming search algorithm may arrive at a point like a and will not be able to distinguish it from the global optimum. The search algorithm can check only for local improvements around a, and there may not be any even though a superior point is feasible farther away. The famous Karush–Kuhn–Tucker conditions can be used to verify that a point located by the search algorithm is, indeed, a local optimum, but cannot verify whether any other (possibly

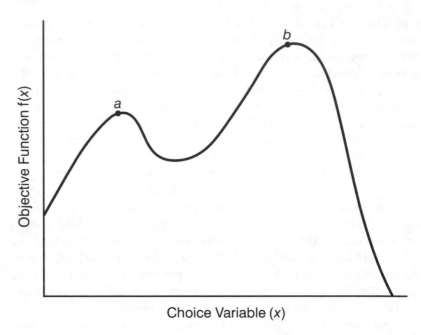

FIGURE 1.1
A nonlinear, nonconvex objective function.

better) local optima exist. The possibility of multiple local optima arises only if the problem is not convex (see Gill et al., 1981) but, unlike the case in figure 1.1, model convexity can be very difficult to confirm or deny. The usual procedure is to use a number of different starting points to lead the algorithm to different local optima if they exist.

The global optimization problem does not seem likely to be remedied in the near future. Even finding local optima is not always easy with nonlinear programming. It is necessary to formulate the model carefully with regard to scaling of parameters and with regard to intermediate calculations so as to avoid numeric problems in the search process.

A Final Introductory Note

These difficulties with integer and nonlinear programs are not all bad. To the degree that the purpose of modeling is to provide a thinking tool, these difficulties force the analyst to experiment with and investigate the structure of the model being built. It could also be argued that because these types of models directly account for system nonlinearities, thresholds, and interactions, they are a more direct thinking analogue. Even if the reader never attempts to solve the formulations in this book, it is hoped that they provide insights into the spatial nature of managing ecosystems.

PART I

STATIC SPATIAL RELATIONSHIPS

As stated in chapter 1, this section of the book is devoted to capturing the basic, static, spatial relationships that reflect the importance of how close things are to each other, what shape things are, the difference between many small things and few large things, and so forth. Because this is a book about managed ecosystems, we begin by looking at the traditional natural resource optimization approaches using linear programming.

Linear programming has become one of the most common analysis techniques in renewable natural resource management and planning (for example, see Wagner 1975 for a general treatment of mathematical programming and Dykstra 1984 for natural resource applications). The intrinsic linearity of the approach is clearly a limitation, but many nonlinear ecosystem responses can be approximated in linear models, as discussed in parts III and IV. The linearity assumptions do cause some problems, however. The most important of these problems is that of accounting for the impact of the spatial configuration of a management action on the outputs of interest. Typically, natural resource management prescriptions are defined on a per acre basis with fixed per acre production coefficients, and the linear program determines the number of acres allocated to each management prescription. The problem is that the nonlinear response to different sizes, shapes, and configurations of a management action cannot be captured with the simple per acre production coefficient.

In terms of ecosystem response, it is often just as important how a management action (such as a timber harvest) is spatially laid out as how many acres are involved (Harris, 1984; Saunders et al., 1991; Franklin and Forman, 1987; Diamond, 1975). Some natural resource applications attempt to deal with this problem by making the linear program behave more like an integer program by defining management prescriptions not on a per acre or stand basis but applying to an entire watershed (or similar land unit), with a predetermined spatial configuration. This allows the prediction of the resulting effects to take a given spatial layout into account, but a serious problem remains. Within the practical limits of building and solving the linear program, only a tiny number of the possible spatial layouts can be considered. An example might be useful to illustrate the point.

Imagine a square watershed. Assume that a low level of resolution is required for estimating the effects of different spatial configurations of vegetative cover, so we segment the watershed into 25 equal-sized cells. Also assume that we have only one management action (such as clearcutting) to consider for each cell, versus doing nothing. Even in this very simple example, if each of the 25 cells could be clearcut or not, 2^{25}, or over 33 million spatial configurations would be possible. If a binary decision variable must be included in an optimization model for each spatial configuration, model size would be prohibitive. Yet if the effects of spatial configuration can be directly estimated within an optimization model, the same watershed problem could be modeled using only 25 binary decision variables (each representing harvest of a particular cell). With more potential management actions or with finer resolution, the number of spatial possibilities approaches googol numbers very rapidly. Add scheduling (time) considerations and the problem gets larger still. Thus, we have a powerful motivation to approximate spatial effects more directly within management optimization models.

In chapters 2 and 3, we discuss basic methods, using cellular and geometric formulations, for capturing the spatial relationships between choice variables, as opposed to containing them within each choice variable. We use static wildlife examples in these chapters, but the approaches set up other applications, as discussed in parts II and III. By approaching the problem in this manner, we directly model the spatial relationships, and the number of choice variables is proportional to the spatial resolution included. In the cellular model, we define spatial resolution by the number of cells included in the grid. In the geometric model, we define spatial resolution by the number of geometric patches of some landscape feature (such as habitat) included. The problems are still difficult to formulate and solve, often requiring either integer or nonlinear programming methods. The cellular approach can accommodate a nonhomogeneous landscape and is solvable with integer methods, thus avoiding local optima. The geometric approach requires a homogeneous landscape and nonlinear methods, but has greater geometric flexibility in some respects, because it is not limited to the cellular grid.

In chapter 4, a different type of basic spatial relationship is discussed, one that results from economic entities (supply and demand) that equilibrate according to their spatial properties. We use a recreation example, which involves the human factors of demand concentrated in population centers and the natural resource factors of spatially located recreation opportunities. Chapter 4 is the only part of the book that emphasizes economic efficiency. Any of the methods in parts I-III could be combined with economic objective functions and thus emphasize economic efficiency while capturing the spatial functioning of the ecosystems involved. In contrast, part IV is devoted to exploring noneconomic objective functions. In chapters 2 and 3, we use very simple weighted-sum objective functions with a biophysical production definition to focus attention on the spatial relationships discussed. After part I discusses basic static spatial relationships, we are ready to move on to stochastic spatial relationships and dynamic spatial relationships in parts II and III, respectively.

Chapter 2

A CELLULAR MODEL OF WILDLIFE HABITAT
SPATIAL RELATIONSHIPS

This chapter focuses on cellular approaches to land allocation modeling that optimize spatial layout for a single time period and use mixed-integer linear programs such that the number of integer choice variables increases proportionally with the level of spatial resolution (see Hof and Joyce, 1993). Wildlife is the primary spatial concern for this discussion. Although wildlife habitat requirements include many factors, we address a subset of these habitat needs related to spatial configuration (edge, fragmentation, and size thresholds). In this situation, choice variables (management actions) have semipermanent impacts on the spatial layout of wildlife habitat, and the problem is to design that layout so that wildlife can effectively use the habitat. The static formulations developed are reasonably powerful when the spatial layout created by the choice variables is long lasting and the static characterization of spatial layout is important. Recent issues involving old growth harvesting are examples involving such semipermanent character changes, and we will develop an example along these lines. It is assumed that the single time period included reflects the permanence of management, and that wildlife populate the habitat to the degree that the spatial layout (edge, fragmentation, and size thresholds) permits. This would be something of an equilibrium population pattern based on that spatial layout. In this sta-

Much of the material in this chapter was adapted from J. G. Hof and L. A. Joyce, A mixed integer linear programming approach for spatially optimizing wildlife and timber in managed forest ecosystems, *Forest Science* 39(4) (1993): 816–834, with permission from the publisher, the Society of American Foresters.

1	2	3	4	5
6	7	8	9	10
11	12	13	14	15
16	17	18	19	20
21	22	23	24	25

FIGURE 2.1
A 25-cell planning area.

tic formulation, dynamic processes such as population movement, births, and deaths are not modeled directly.

We begin by dividing the land area into cells, as depicted in figure 2.1. Choice variables, C_i, are defined such that $C_i = 0$ if the ith cell is harvested and $C_i = 1$ if it is left in old growth. Let us assume that we wish to maximize the weighted sum of the expected population of a species that prefers recently harvested areas (which could also represent a timber output), the expected population of a species that is dependent on the area of a preharvest habitat (old growth), and the expected population of a species that is determined as a linear function of edge (E). Thus we would maximize

$$\sum_{k=1}^{3} V_k S_k, \tag{2.1}$$

where

V_k = the objective function weight for the kth species, reflecting the relative importance of that population in selecting an overall management strategy,

S_1 = expected population of the young-forest–dependent, species 1,

S_2 = expected population of an area-dependent, species 2, and

S_3 = expected population of an edge-dependent, species 3.

Also, assume that species 1 output can be calculated simply by the following constraint:

$$S_1 = \sum_{i=1}^{M} t_i(1 - C_i),\qquad(2.2)$$

where

t_i = the expected species 1 population from the ith cell, if harvested, and

M = the number of cells in the land area.

The remaining problems are thus to determine the amount of S_2 and S_3 from the choice variables, C_i. The next section discusses the calculation of edge (E), and it is then assumed that S_3 is a linear function of edge:

$$S_3 = gE,\qquad(2.3)$$

where

g = the expected number of species 3 from each kilometer of edge (assumed to be 1.26157 hereafter, based on assumptions in Hof and Joyce, 1992, for goshawk as an edge-dependent).

The following section then discusses the calculation of S_2 so as to account for fragmentation effects. A size threshold on habitat area is also formulated.

Edge Effects

The species 3 population is assumed to be determined by the amount of edge between a mature stand of timber and a cutover area. Such a species is called an edge-dependent here. An edge-dependent might be a small mammal that cannot range far and has habitat requirements involving both the old growth and cutover areas or a species (such as the goshawk) that preys on those small mammals. A nonlinear approach for calculating the amount of edge is

$$N = \sum_{i=1}^{M} C_i,\qquad(2.4)$$

$$E = N(4 \cdot L) - L \cdot \left[\sum_{i=1}^{M} C_i \left(\sum_{j=1}^{M} \delta_{ij} C_j \right) \right],\qquad(2.5)$$

where

N = the number of cells left in old growth,

E = the amount of edge between old growth and harvested areas,

L = the length of one cell's side, and

δ_{ij} = 1 if cell i shares a side with j, 0 otherwise.

Constraint (2.4) calculates the total number of cells left in old growth. It is assumed (for simplicity) that the land around the planning area is already cut over, so that when the C_i values for the border cells equal 1, the outer perimeter is edge. Constraint (2.5) could easily be modified for different conditions. Constraint (2.5) calculates the amount of edge as follows. The amount of edge from each C_i that is equal to 1 is $(4 \times L)$ if all adjacent C_i are 0. Thus constraint (2.5) starts with $N(4 \times L)$. For each adjacent C_i that is not 0, the edge is overestimated by $(2 \times L)$. Constraint (2.5) thus deducts the overestimates accordingly. This approach is nonlinear, but it was found in Hof and Joyce (1992) that, without further complications, it solved naturally with an integer solution if the C_i were continuous variables constrained to be between 0 and 1. In the absence of a mixed-integer nonlinear solution procedure, further elaboration of this approach seems impractical.

In fact, it is possible to reformulate this basic approach so that it is linear, and thus solvable by mixed-integer linear methods. This formulation is

$$N = \sum_{i=1}^{M} C_i, \tag{2.6}$$

$$E = N(4 \cdot L) - L\left(\sum_{i=1}^{M} D_i\right), \tag{2.7}$$

$$D_i \leq \alpha_i C_i \qquad \forall\, i, \tag{2.8}$$

$$D_i \geq \left(\sum_{j=1}^{M} \delta_{ij} C_j\right) - \alpha_i(1 - C_i) \qquad \forall\, i, \tag{2.9}$$

where

C_i = an integer variable that is 0 if cell i is harvested and 1 if cell i is left in old growth,

D_i = a continuous, nonnegative accessory variable for each i, which indicates the number of deductions (of L) taken for each i in calculating E, and

α_i = an arbitrarily large constant (such that $\alpha_i \geq \sum_{j=1}^{M} \delta_{ij} C_j \quad \forall\, i$).

Other variables are as defined before.

Each pair of equations such as

$$D_1 \leq \alpha_1 C_1, \tag{2.10}$$

$$D_1 \leq \sum_{j=1}^{M} \delta_{1j} C_j - \alpha_1 (1 - C_1), \tag{2.11}$$

taken with equations (2.6) and (2.7), serves the purpose of deducting the edge overestimates. For example, if $C_1 = 0$, then equation (2.10) forces D_1 to be 0 (no deductions) and equation (2.11) is "switched off" because $1 - C_1 = 1$. When $C_1 = 1$, then equation (2.10) is "switched off" and equation (2.11) becomes

$$D_1 \geq \sum_{j=1}^{M} \delta_{1j} C_j. \tag{2.12}$$

Because D_1 is deducted from the edge calculation and E is positively valued (through g) in the maximization objective function, D_1 solves equal to $\sum \delta_{1j} C_j$ and the correct amount of edge overestimate is deducted.

This linear approach in equations (2.7)–(2.9) requires many more constraints than equation (2.5), but more importantly, the number of integer variables equals the number of cells in the grid. The formulations that follow retain the property that the number of integer choice variables is proportional to the number of cells, so that fairly large problems might be solvable (at least to a satisficing solution).

It should also be noted that in the geometric approach discussed in chapter 3, we include an annulus of forage around each old growth area that serves as the effective feeding area around the cover area for animals such as deer. In the formulation in equations (2.6)–(2.9), this feeding area could be approximated as a linear function of edge. For example, for every kilometer of edge, there could be 1/3 km² of feeding area within 1/3-km of the old growth.

Wildlife Habitat Fragmentation Effects

Sessions (1992) has discussed habitat connections as a network or tree where habitat is considered nonfragmented if some path or corridor exists that connects all habitat areas. Sessions then focuses on finding the minimum-length path that connects all previously located critical habitat areas. This approach has its appeal and is consistent with the movement corridor concept (Harris and Gallagher, 1989; see also Simberloff et al., 1992, for an alternative viewpoint). The problem addressed in this chapter, however, is to optimally locate wildlife habitat and harvested areas in the first place. Also, treating fragmentation as a connected graph or network problem seems to treat wildlife habitat connectivity as systematic, when, in fact, it may be random.

The static concept of habitat connectivity should reflect the long-term movement behavior of wildlife, even though we are not modeling wildlife population movement (or dynamics) per se. Considerable theoretical ecology literature suggests that wildlife dispersal is random in nature (for example, see Skellam, 1951; Okubo, 1980). If dynamic wildlife movements are random, this suggests that the static condition of habitat connectivity is also random. Thus if a group of areas is not perfectly conjoined with corridors, the probability of all areas being populated in any amount of time is diminished, but not necessarily 0 (assuming the species can move through nonhabitat to some degree). Likewise, even a habitat system perfectly conjoined with corridors may not have a 100 percent probability of being totally populated within any given amount of time. We define the static condition of connectivity of habitat areas as a probabilistic condition that reflects equilibrium dispersal patterns, given the permanence of management actions (which define the single time period for the analysis in the first place). The ecological literature typically indicates that habitat areas grouped, for example, in a triangle or square are preferable to those grouped in a row (see Diamond, 1975; Soulé, 1991). Likewise, a circle is a better shape than an oblong (Game, 1980), and one big area is better than several smaller connected ones (Simberloff, 1988). These preferences are based on a notion that many wildlife populations radiate in a more or less directionless (360°) fashion, so clumped configurations are easier to populate. This distinction would be lost in a movement-corridor analysis.

The directionless (360°) dispersal of wildlife applies to expected values, and typically involves a distance–decay relationship. Put another way, there is some probability that a given habitat area near any other habitat area is connected, and this probability diminishes as the distance increases between the two habitat areas. With many habitat areas, the probability of a given area being connected to the group is a function of the number of other habitat areas nearby and the distances to them. We assume that the probability of each area being connected to a group of areas is the joint probability that the area is connected to *any* (not all) of the areas in the group. We also assume independence between the individual connectivity probabilities. Thus, the joint probability (PR_i) of each cell i being connected is

$$PR_i = 1 - \left[\prod_{j=1}^{M}(1 - pr_{ij}C_i)\right] \quad \forall i \qquad (2.13)$$

$$(pr_{ii} = 0),$$

where pr_{ij} is the probability that cell i is connected to cell j. Presumably, pr_{ij} would be smaller the further cell j is from cell i. Equation (2.13) simply calculates the joint probability that cell i is not connected to any of the $j = 1, \ldots, M$ ($j \neq i$) cells, and then calculates PR_i as the converse of that joint probability. Looked at this way, wildlife habitat connectivity is similar to a gravity model: An area has a very high probability of being connected if it is surrounded closely by other areas of habitat, and a lower probability with fewer surrounding habitat areas or with greater distances between areas. At some distance, the probability of two habitat areas being connected is effectively 0. Thus, when an area is left as habitat, it has a certain probability of being connected, which is determined by the number and location of other habitat areas, and it also contributes to the probability of other areas being connected in an equivalent manner. Cells of habitat contribute to supporting populations only to the degree that they are probabilistically connected.

We can approximate equation (2.13) with a mixed-integer linear set of constraints. Let us assume that the probability of a cell being connected to an adjacent cell is 0.5, that the probability of it being connected to the next one over is 0.15, and that there is 0 probability of connectivity if the cell is more than one cell removed. For convenience, define the set Ω_i as the indexes of the cells that immediately surround the ith cell, and the set θ_i as the indexes of the cells that surround Ω_i. Because the planning area is surrounded by cut-over area and we will use Ω_i and θ_i for habitat connectivity, we will limit Ω_i and θ_i to the planning area. Equation (2.13) could then be approximated by

$$PR_i \leq \sum_{j \in \Omega_i} .3C_j + \sum_{k \in \theta_i} .1C_k \quad \forall i, \qquad (2.14)$$

$$PR_i \leq C_i \quad \forall i. \qquad (2.15)$$

The coefficients (.3 and .1) were chosen to approximate equation (2.13) with the probabilities of .5 and .15 for the two sets of surrounding cells. Constraint (2.15) prevents PR_i from exceeding 1 (because $C_i \leq 1$). Also, if $C_i = 0$ in solution, then PR_i is forced to 0 by equation (2.15).

We apply this approach to species 2, the old growth area-dependent. The expected value of the total species 2 population, S_2, would thus be

$$S_2 = \sum_{i=1}^{M} \text{PR}_i \cdot p_i, \qquad (2.16)$$

where p_i is the expected species 2 yield from C_i if $C_i = 1$ and $\text{PR}_i = 1$.[1] The S_2 in the objective function is thus determined by equations (2.14)–(2.16).

If only the sets Ω_i are relevant (the probabilities are 0 for all θ_i) then a very accurate approximation of equation (2.13) could be accomplished by replacing equation (2.14) with

$$Q_i = \sum_{j \in \Omega_i} C_j \qquad \forall\, i, \qquad (2.17)$$

$$Q_i \geq \sum_{k=1}^{6} X_{ik} \qquad \forall\, i, \qquad (2.18)$$

$$X_{ik} \leq 1 \qquad \forall\, i \qquad (2.19)$$
$$\text{for } k = 1, \ldots, 5,$$

$$\text{PR}_i \leq .5X_{i1} + .25X_{i2} + .125X_{i3} + .0625X_{i4}$$
$$+ .03125X_{i5} + .015625X_{i6} \qquad \forall\, i. \qquad (2.20)$$

It is still important to include equation (2.15), of course. Notice that this approach adds about $2 \cdot M$ rows to the formulation, plus $5 \cdot M$ upper bounds (2.19), but does not require any additional integer variables. This approach is exact, except that the seventh adjacent old growth cell makes PR_i equal to 1 (it is correct at .984375 after six adjacent cells), whereas in equation (2.13), PR_i approaches 1 asymptotically and is .9921875 after seven cells. Any degree of accuracy could be achieved by adding additional segments (X_{ik}) to the formulation (which might be very desirable with pairwise probabilities smaller than 0.5). The approach in equation

---

1. The cells of the landscape have different quality and ability to support wildlife. The expected number of animals that a cell supports reflects the probability of the cell being connected to other cells and the potential number of animals that the cell could support. The model emphasizes aggregation of habitat rather than the establishment of distant but suitable habitats. This formulation applies to animals that colonize areas near other animals rather than in distant but vacant patches (Weddel, 1991; Milne et al., 1989).

(2.14) is compared to the approach in equations (2.17)–(2.20) for the case that considers only the Ω_i set of surrounding cells in this chapter's appendix. It is concluded that the linear approximation in equation (2.14) is effective, and it is used throughout the remainder of this chapter to allow the (approximated) inclusion of the θ_i sets of surrounding cells.

Wildlife Habitat Size Thresholds

An area-dependent animal might very well have a minimum habitat size requirement, below which no population is maintained. This can be formulated for species 2 by adding

$$S_2 \leq \gamma \cdot R, \tag{2.21}$$

$$R \leq (1/\text{TH}) \cdot N, \tag{2.22}$$

where
R = a 0–1 integer variable,
TH = the size threshold in cell units (number of cells needed to meet the minimum size requirement),
γ = an arbitrarily large constant (set so γ exceeds the maximum possible value of S_2),

and N is as previously defined. If N is less than TH then R is forced by equation (2.22) to be 0, and S_2 is forced by equation (2.21) to be 0. On the other hand, if N is greater than or equal to TH then R can equal 1, which makes equation (2.21) nonbinding on S_2. Note that equations (2.21) and (2.22) together require one additional integer variable (R). Habitat selection has been theorized at different scales—optimal foraging, patch movement and selection, immigration and migration—with similar theoretical decision rules across these different scales (Orians, 1991). In this chapter, we examine a habitat size threshold for population viability, but these constructs could be applied at a variety of scales. In the next section, we present a simple example to demonstrate these formulations.

An Example

The Problem

Suppose we wish to plan for an area as mapped in figure 2.2a. This area has varied topography that is generally flatter in the southern 3/5, resulting in higher timber yield. A river meanders through the area. A hiking

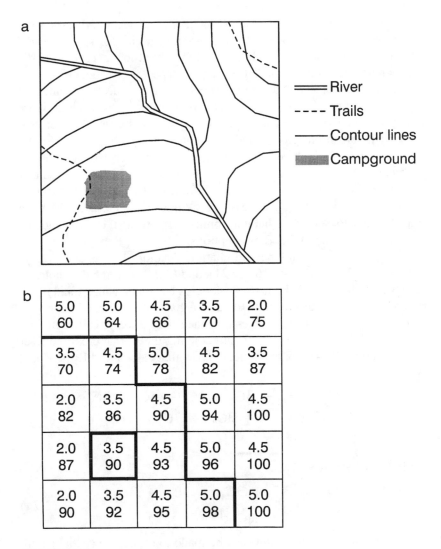

FIGURE 2.2
A mapped planning area (a) and its digitized equivalent (b).

trail accesses a campground in the southwest quadrant, and another trail passes through the northeast corner. It is assumed that the hiking trails degrade the quality of wildlife habitat, but that they do not affect wildlife habitat connectivity. For demonstration purposes, we make two different assumptions regarding the impact of the river and the campground: first, that neither affects wildlife habitat (Model A), and second, that the river completely negates wildlife habitat area connectivity for species 2 and that the campground area yields no wildlife habitat for species 1 or 2 but acts as harvested area in calculating edge for species 3 (Model B).

Figure 2.2b depicts a "digitized" version of the area. We assume that each cell is 3 km × 3 km, so $L = 3$. The top number in each cell in figure 2.2b gives the expected cellular population (p_i) of species 2, given that the area is left in old growth and connected with probability of 1. The bottom number in each cell gives the expected species 1 population for each cell (t_i), if it is harvested. These numbers would be the result of topography, disturbances such as the hiking trails, and any number of other site quality factors. The heavy black lines in figure 2.2b show the impediments to species 2 habitat connectivity from the river and the campground for the Model B assumptions.

Model A includes all the formulation developed: equations (2.1)–(2.3), (2.6)–(2.9), (2.14)–(2.16), (2.21), and (2.22) with the threshold set at two cells. Model B is built from Model A by: modifying t_{17} and p_{17} to 0, forcing $C_{17} = 0$ to contribute to edge as harvested area and to prevent it from contributing to the species 2 connectivity of other cells (but with species 1 yield, t_{17}, still set at 0), and deleting the C_i that cross the river from equation (2.14). For example, the row that calculates PR$_7$ in Model A would be

$$PR_7 = .3C_1 + .3C_2 + .3C_3 + .1C_4 + .3C_6 + .3C_8 \qquad (2.23)$$
$$+ .1C_9 + .3C_{11} + .3C_{12} + .3C_{13} + .1C_{14}$$
$$+ .1C_{16} + .1C_{17} + .1C_{18} + .1C_{19},$$

but in Model B it would be

$$PR_7 = .3C_6 + .3C_{11} + .3C_{12} + .3C_{13}$$
$$+ .1C_{16} + .1C_{17} + .1C_{18}. \qquad (2.24)$$

In addition, other modifications can be made to reflect the spatial nature of the river. For example, in calculating PR$_1$, we eliminated not only C_6, C_7, C_{11}, C_{12}, and C_{13}, but also C_8 because the route (around the river) between cell 8 and cell 1 is so circuitous. This was the only modification of this type that we made.

Results

Figure 2.3 presents the spatial solution for Model A (figure 2.3a) and Model B (figure 2.3b) with $V_1 = 10$, $V_2 = 300$, and $V_3 = 20$. Tables 2.1 and 2.2 present the numerical solutions for all the figures. In all the fig-

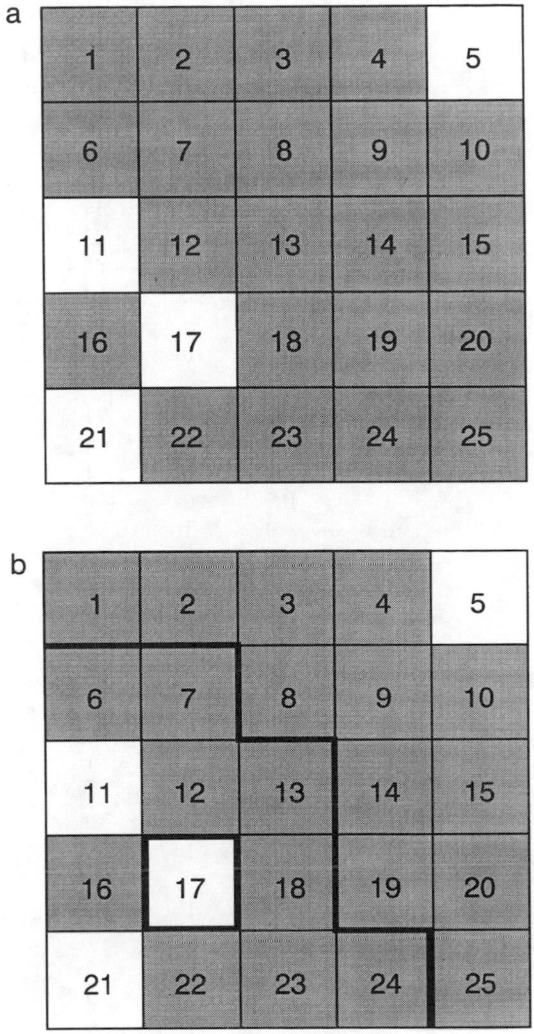

FIGURE 2.3
Solutions to Models A and B with $V_1 = 10$, $V_2 = 300$, and $V_3 = 20$.

ures, the shaded area indicates old growth ($C_i = 1$). The objective function weight on species 1 is relatively low, so the results in figure 2.3 leave most of the area in old growth. The harvesting that is included is arranged so as to provide as much edge as possible for species 3. With this emphasis on old growth wildlife habitat, the spatial solutions for

TABLE 2.1
Solutions for Figures 2.3–2.5.

	Figure 2.3		Figure 2.4		Figure 2.5	
	a	b	a	b	a	b
Objective function	32,488	30,133	35,812	32,697	34,080	31,575
S_1	337	247	886	1219	1252	1575
S_2	90.5	85.6	67.0	43.5	51.0	26.5
S_3	98	98.4	121.1	68.1	0	0
Edge	78	78	96	54	—[a]	—[a]
Total old growth cells	21	21	15	10	11	6
Total harvested cells	4	3	10	14	14	18
PR_1	1	.4	1	.4	1	0
PR_2	1	1	1	1	1	.9
PR_3	1	1	1	1	1	1
PR_4	1	1	1	1	0	1
PR_5	0	0	0	0	0	0
PR_6	1	.9	1	0	1	0
PR_7	1	1	1	0	1	0
PR_8	1	1	1	1	1	1
PR_9	1	1	1	1	1	1
PR_{10}	1	1	0	0	0	0
PR_{11}	0	0	0	0	0	0
PR_{12}	1	1	0	0	0	0
PR_{13}	1	1	1	0	1	0
PR_{14}	1	1	0	1	1	.9
PR_{15}	1	1	1	1	0	0
PR_{16}	1	1	0	0	0	0
PR_{17}	0	0	1	0	0	0
PR_{18}	1	1	0	0	1	0
PR_{19}	1	1	1	1	1	0
PR_{20}	1	1	0	1	0	0
PR_{21}	0	0	0	0	0	0
PR_{22}	1	1	0	0	0	0
PR_{23}	1	1	1	0	0	0
PR_{24}	1	.9	1	0	0	0
PR_{25}	1	.8	.9	0	0	0

[a] Not valued.

Models A and B in figure 2.3 are the same. However, from table 2.1, it is clear that the expected value of the species 2 population is lower in Model B because of the reduction in connectivity. In fact, because the river is impassable in Model B, there are actually two unconnected habitat areas created in the Model B solution. Each area is still well-

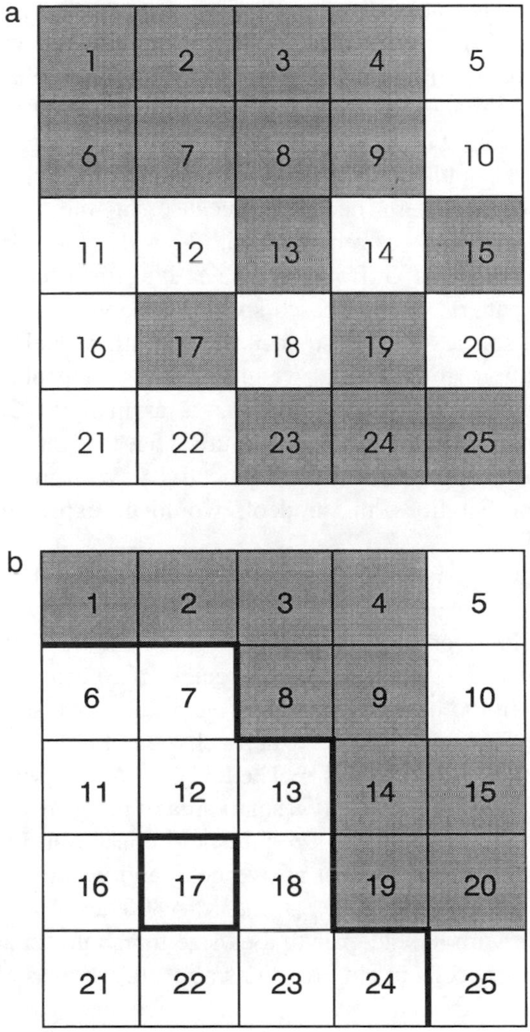

FIGURE 2.4
Solutions to Models A and B with $V_1 = 15$, $V_2 = 300$, and $V_3 = 20$.

connected within itself, however. The species 1 population in Model B is also lower because of the campground.

Figure 2.4 was generated with $V_1 = 15$, $V_2 = 300$, and $V_3 = 20$. The higher species 1 weight causes more harvesting in solution. In figure 2.4b, the effect of the river is quite pronounced. In contrast to figure

2.3b, there is only one well-connected old growth area in figure 2.4b. And, from table 2.1 it is clear that the river impediment significantly reduces the expected population of species 2 in solution relative to figure 2.4a. In 2.4a, the old growth is spread out through most of the area and is laid out in something of a checkerboard in the southern part of the area, where species 1 population coefficients are higher. We hypothesized that the checkerboard pattern occurred because of the edge-dependent species 3, so figure 2.5 was generated with $V_1 = 15$, $V_2 = 300$, and $V_3 = 0$. Eliminating species 3 from the objective function removed the checkerboard pattern, and increased species 1 population somewhat, as expected. The species 1 increase occurred in both Models A and B because the solutions in figure 2.4 weigh both edge and old growth area against species 1, not just old growth area, as in figure 2.5. It is also clear, in comparing figure 2.4a with figure 2.5a, that there is a tradeoff between edge and old growth area connectivity. With more pessimistic connectivity probabilities, this tradeoff would be expected to be even more pronounced.

Figure 2.6 was generated with $V_1 = 20$, $V_2 = 300$, and $V_3 = 20$. With this emphasis on species 1, only a small area of old growth is retained. The effect of the river in figure 2.6b is to move the five-cell old growth area over one cell to the east. The species 2 expected population is slightly lower in the Model B solution (table 2.2) because the species 2 population coefficients in the relocated cells are slightly lower than the ones left in old growth in Model A. The level of connectivity is identical, however. This solution, with only a small area of old growth, provides an opportunity to demonstrate the size threshold constraint. In all previous solutions, this threshold was set at two cells and was thus nonbinding. Figure 2.7 was generated with $V_1 = 20$, $V_2 = 300$, and $V_3 = 20$ (same as figure 2.6), but with the old growth area size threshold set at eight cells. Model A left eight cells in old growth as a result, whereas Model B harvested the entire area; the river caused the connectivity and wildlife yield of eight old growth cells to be less desirable in terms of objective function attainment than complete harvesting. This demonstrates that the size threshold can cause the solution to be "all-or-nothing." It also shows that landscape features can strongly affect the solution.

Discussion

The solutions in figures 2.3–2.7 and tables 2.1 and 2.2 seem to respond reasonably to the parameter changes. Overall model tenability probably rests on empirical questions regarding wildlife behavior that are not cur-

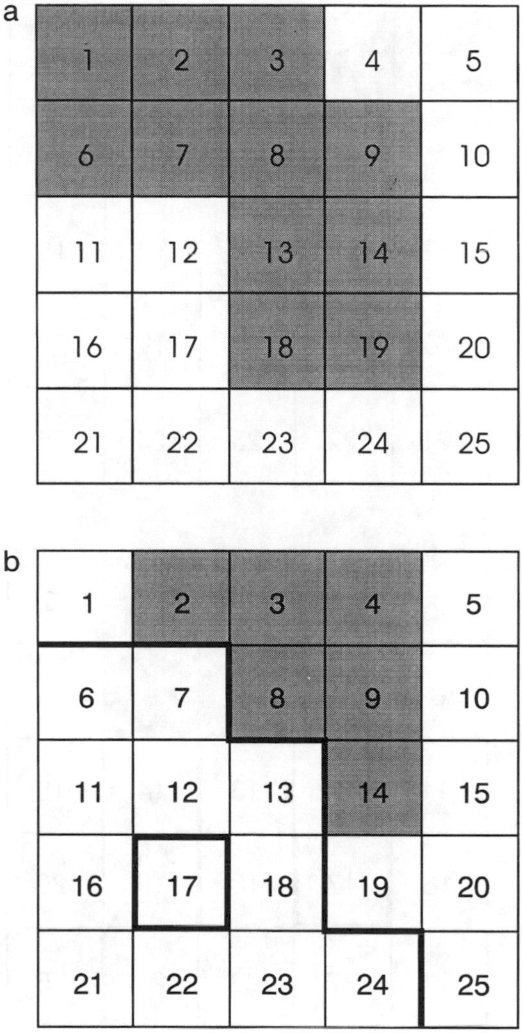

FIGURE 2.5
Solutions to Models A and B with $V_1 = 15$, $V_2 = 300$, and $V_3 = 0$.

rently well understood and that may vary from case to case. Nonetheless, the approach is based on a logical construction (gravity model) that has proved to be empirically useful in a number of settings.

Despite its integer variables, the model solves reliably. Without the edge effects, the model solves very rapidly (around 10 seconds clock

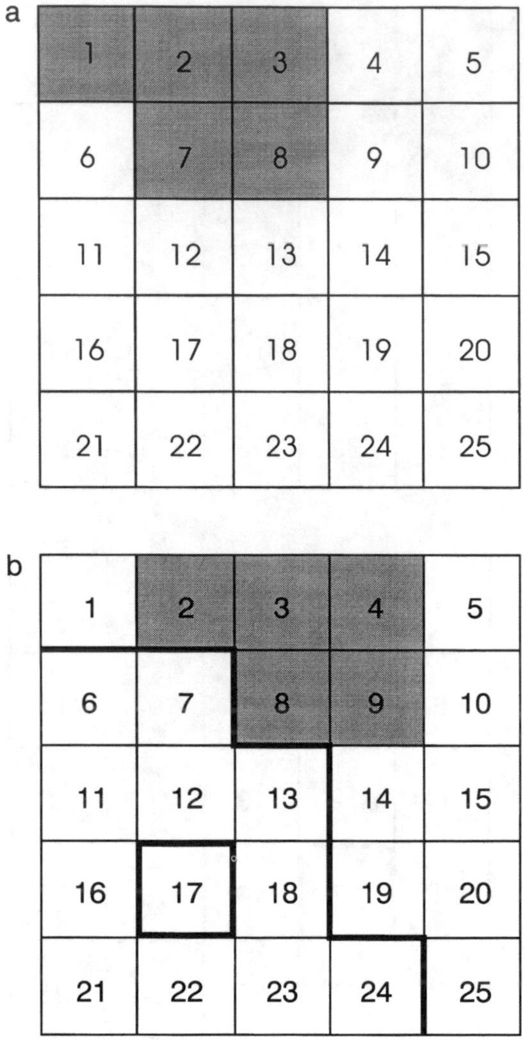

FIGURE 2.6
Solutions to Models A and B with $V_1 = 20$, $V_2 = 300$, and $V_3 = 20$.

time with a 486/33 microcomputer). The solutions in figures 2.3–2.7 (with edge effects) had clock times more in the range of $\frac{1}{2}$ to 2 hours. As hardware and software for these kinds of problems continue to improve, many more cells could be included for improved spatial detail (in fact, it appears that many more could be included at this time).

TABLE 2.2
Solutions for Figures 2.6 and 2.7

	Figure 2.6		Figure 2.7	
	a	b	a	b
Objective function	43,196	40,587	42,710	40,580
S_1	1777	1669	1535	2029
S_2	23.0	21.5	36.5	0
S_3	37.8	37.8	53.0	0
Edge	30	30	42	0
Total old growth cells	5	5	8	0
Total harvested cells	20	19	17	24
PR_1	.8	0	1	0
PR_2	1	.8	1	0
PR_3	1	1	1	0
PR_4	0	1	0	0
PR_5	0	0	0	0
PR_6	0	0	1	0
PR_7	1	0	1	0
PR_8	1	1	1	0
PR_9	0	1	1	0
PR_{10}	0	0	0	0
PR_{11}	0	0	0	0
PR_{12}	0	0	0	0
PR_{13}	0	0	1	0
PR_{14}	0	0	0	0
PR_{15}	0	0	0	0
PR_{16}	0	0	0	0
PR_{17}	0	0	0	0
PR_{18}	0	0	0	0
PR_{19}	0	0	0	0
PR_{20}	0	0	0	0
PR_{21}	0	0	0	0
PR_{22}	0	0	0	0
PR_{23}	0	0	0	0
PR_{24}	0	0	0	0
PR_{25}	0	0	0	0

Appendix

The formulation in equations (2.17)–(2.20) applies only when the probability of a cell being connected to other cells is the same (or 0) for all those other cells. Thus, the linear approximation in equation (2.14) can be compared to (2.17)–(2.20) when only the Ω_i set of surrounding cells is included for each i (with an identical probability associated with all

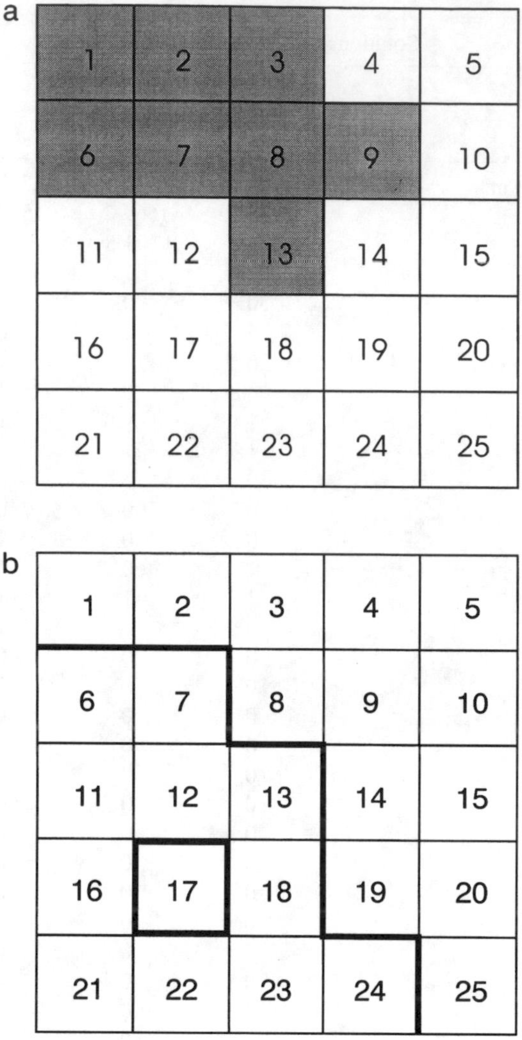

FIGURE 2.7
Solutions to Models A and B with $V_1 = 20$, $V_2 = 300$, $V_3 = 20$, and a species 2 size threshold of eight cells.

TABLE 2.3

Comparison of Solutions from the Very Accurate Formulation in Equations
(2.17)–(2.20) with Those from the Linear Approximation in Equation (2.14)

	Species 1 Weight	Objective Function Value	Species 1 Expected Population	Species 2 Expected Population	Number of Different Cells in Solution
Solutions with	5	29,044	0	96.8	0
equations	10	29,925	334	88.6	0
(2.17)–(2.20)	15	33,214	1441	38.7	2
	18	38,277	1777	21.0	1
	20	42,380	2119	0	0
Solutions with	5	29,580	0	98.6	0
equation (2.14)	10	30,535	334	90.7	0
	15	33,675	1252	49.7	2
	18	38,436	1707	25.7	1
	20	42,380	2119	0	0
Solutions with	5	29,044	0	96.8	—
equation (2.14)	10	29,925	334	88.6	—
imposed on formu-	15	33,077	1252	47.7	—
lation with equa-	18	38,212	1707	25.0	—
tions (2.17)–(2.20)	20	42,380	2119	0	—

members of Ω_i). Table 2.3 presents such a comparison. The edge-dependent species 3 (including equations 2.7–2.9) and the size threshold (equations 2.21 and 2.22) were deleted to avoid obfuscation of results. The objective function weight for species 2 was held constant at 300, and the weight for species 1 was varied from 5 to 20 to vary the output emphasis from one extreme to the other. The top set of solutions was generated with a formulation that used (2.17)–(2.20), and is a very accurate approximation to the true optimum. The middle set of solutions was generated with a formulation that used (2.14) as a linear approximation. The bottom set of solutions was generated by imposing the middle solutions on the formulation that used (2.17)–(2.20). This bottom set of solutions indicates the actual output and objective function values that would result from the middle solutions.

The objective function values compare rather favorably. The solutions with (2.14) are never more than 0.5 percent suboptimal, based on comparing the bottom set of values with the top. The output emphases are only slightly different between the accurate and approximated models.

This is not surprising because, as indicated by the far right column in table 2.3, the solution values for the cellular choice variables (C_i) never varied in more than two cells.

Overall, for this case, the linear approximation (2.14) appears to be effective. The main advantage of the approach in (2.14) is that it allows the inclusion of cells that are different distances from each other (such as set Ω_i for each $_i$). We cannot really test for the reliability of optimization results for this case, because equations (2.17)–(2.20) are not applicable. Nonetheless, to the degree that the comparison in table 2.3 applies, the approach in (2.14) appears to be reliable relative to the overall expected accuracy in the data for an analysis such as this. For the purposes of this chapter, the inclusion of cells such as those in the θ_i sets is important, so (2.14) is used throughout the case example. In other contexts, the accuracy of (2.17)–(2.20) may be more important.

Chapter 3

A GEOMETRIC MODEL OF WILDLIFE HABITAT SPATIAL RELATIONSHIPS

The approach discussed in this chapter is called a geometric structure because it uses geometric shapes to characterize the spatial layout of vegetation types resulting from management actions such as timber harvesting (see Hof and Joyce, 1992). In this chapter, we use circles because of various conveniences, but other shapes (squares, rectangles, ellipses, etc.) could also be used (and are used in chapter 6). Irregular shapes can be approximated by a set of regular shapes. For example, an S-shaped area could be approximated by a string of appropriately located circles. If we define protected areas as circles (or cutover areas could be so defined), then the choice variables become the size and location of these circles. It is necessary to prespecify the maximum possible number of circles, but the approach is designed so that fewer numbers of circles can be selected in the optimization by making the size of some circles 0. We can choose the location of the circles by defining a system of rectangular (x,y) coordinates for their centers, and we can choose the size of the circles by determining their radii. To demonstrate this approach, we again examine a wildlife habitat spatial allocation problem similar to the problem in chapter 2.

Some of the material in this chapter was adapted from J. G. Hof and L. A. Joyce, Spatial optimization for wildlife and timber in managed forest ecosystems, *Forest Science* 38(3) (1992): 489–508, with permission from the publisher, the Society of American Foresters. The formulations are actually quite different, however, reflecting the evolution of our approaches.

Spatial Effects

As in chapter 2, we postulate that edge, habitat fragmentation, and a minimum size threshold for usable patches of habitat all affect populations of different species. In addition, we use the geometric approach to account for the amount of one habitat type that is juxtaposed around another (an extension of the edge effect). An example of this effect might be a species (such as a species of deer) that requires old growth habitat for cover or other purposes, but uses open areas for foraging. Only the forage within a certain distance of the old growth might be used because the animals will not venture farther than that distance from the cover.

It is clear that many species are influenced by more than one of these determinants. A useful (and reasonably tenable) biological approach assumed here is that a given species' population is determined by the most limiting of many factors. Of course, a number of factors might simultaneously limit a given species' population, especially in an optimal solution.

An Example

Having developed the context for static spatial wildlife habitat selection in chapter 2, we go straight to a demonstrative example of the geometric formulation, with parameter values based on those in Hof and Joyce (1992). We assume four species are of concern: three species similar to the ones analyzed in chapter 2, and a fourth species to demonstrate the juxtaposition of multiple habitat types as a population determinant. Species 1 prefers harvested area and is not spatially sensitive. Species 2 is assumed to be an old growth–dependent whose habitat is determined by the area of old growth that is spatially connected. It is assumed, as in chapter 2, that the probability of a given patch being connected is the joint probability that it is connected to any of the other habitat patches in the study area. For this example, simple logistic functions relate distance to pairwise probability of connectivity. Species 3 is assumed to be an edge-dependent whose habitat is determined strictly by the amount of edge. Species 4 is assumed to be a wide-ranging species whose population could be determined by either old growth habitat (for cover) or the open-area forage within .366 km of that old growth, whichever is limiting. The size threshold is applied to this species (it is assumed that no species 4 habitat occurs in any patch with a radius smaller than 0.86 km). For this geometric model, we assume a homogeneous landscape in the study area. Assuming an example management area 7.5 km by 15 km

(112.5 km^2) of old growth forest surrounded by grassland, a geometric model formulation capturing all these effects is as follows:

Maximize

$$\sum_{k=1}^{4} V_k S_k, \tag{3.1}$$

subject to

$$x_i \geq r_i \qquad \forall\, i, \tag{3.2}$$

$$y_i \geq r_i \qquad \forall\, i, \tag{3.3}$$

$$x_i + r_i \leq 7.5 \qquad \forall\, i, \tag{3.4}$$

$$y_i + r_i \leq 15 \qquad \forall\, i, \tag{3.5}$$

$$d_{ij} = \left[(x_i - x_j)^2 + (y_i - y_j)^2\right]^{.5} - (r_i + r_j) \qquad \forall\, i, \quad \forall\, j, \tag{3.6}$$

$$p_i = 6.2831853 \cdot r_i \qquad \forall\, i, \tag{3.7}$$

$$A_i = 3.14159265 \cdot r_i^2 \qquad \forall\, i, \tag{3.8}$$

$$S_1 = 10\left(112.5 - \sum_{i=1}^{4} A_i\right), \tag{3.9}$$

$$S_2 = .426 \sum_{i=1}^{4} PR_i A_i, \tag{3.10}$$

$$PR_i = 1 - \left[\prod_{j \neq i}(1 - pr_{ij})\right] \qquad \forall\, i, \tag{3.11}$$

$$pr_{ij} = 1 - \frac{1}{1 + e^{-d_{ij}/1.5}} \qquad \forall\, i, \quad \forall\, j, \tag{3.12}$$

$$S_3 = 1.26 \cdot \sum_{i=1}^{4} p_i, \tag{3.13}$$

$$S_4^i \leq 41.2 A_i' \qquad \forall\, i, \tag{3.14}$$

$$S_4^i \leq 27.4 \cdot F_i \qquad \forall\, i, \tag{3.15}$$

$$F_i = (2.2996458 \cdot r_i) + .4208352 \qquad \forall\, i, \tag{3.16}$$

$$A_i' = 3.141592654 \cdot (r_i')^2 \qquad \forall\, i, \tag{3.17}$$

$$r_i' = r_i \cdot \frac{.5\left[(r_i - .86)^2\right]^{.5} + .5(r_i - .86)}{r_i - .86} \qquad \forall\, i, \tag{3.18}$$

$$S_4 = \sum_{i=1}^{4} S_4^i, \qquad (3.19)$$

$$r_i \geq 0 \qquad \forall\, i,$$

$$d_{ij} \geq 0 \qquad \forall\, i, \quad \forall\, j,$$

where

x_i = the x-coordinate (horizontal direction) of the center of the ith circle of old growth,

y_i = the y-coordinate (vertical direction) of the center of the ith circle of old growth,

$i, j = 1, \ldots, 4$,

S_1 = the expected number of species 1 animals,

S_2 = the expected number of species 2 animals,

S_3 = the expected number of species 3 animals,

S_4 = the expected number of species 4 animals,

S_4^i = the expected number of species 4 animals in patch i,

V_k = the objective function weight for the kth species $(1, \ldots, 4)$,

d_{ij} = the distance between perimeters of circles i and j,

r_i = the radius of the ith old growth circle,

p_i = the perimeter of the ith old growth circle,

PR_i = the joint probability of the ith cell being connected,

pr_{ij} = the probability that the ith cell is connected to the jth cell,

A_i = the area of the ith old growth circle,

F_i = the amount of forageable open area within .366 km of the ith circle of old growth,

A_i' = the area for species 4 in the ith old growth circle ($A_i' = 0$ when $r_i \leq .86$, $A_i' = A_i$ otherwise), and

r_i' = the radius for calculating A_i' ($r_i' = 0$ when $r_i \leq .86$, $r_i' = r_i$ otherwise).

As noted earlier, the specific parameters are based on those derived in Hof and Joyce (1992).

Equation (3.1) is the objective function. Note that the S_k variables are expected values and the V_k are weights, as in chapter 2. Constraints (3.2)–(3.5) force all old growth circles to be inside the 7.5-km by 15-km management area. Constraint (3.6) calculates the distance between perimeters of all pairs of old growth circles. By constraining all d_{ij} (for all circle pairs) to be nonnegative, overlapping old growth circles are prevented. This is necessary to avoid double counting and to preserve the definition of the circles as mutually exclusive areas of old growth. Constraining the radii to be nonnegative prevents negative circle areas. Constraints (3.7) and (3.8) calculate the perimeter and area of each old growth circle, respectively. For this simple example, we prespecify the maximum possible number of circles to be four.

Constraint (3.9) indicates that 10 units of species 1 (the young-growth area-dependent) population are obtained from each square kilometer of area harvested (not left in old growth). Constraints (3.10)–(3.12) determine the population of species 2 (the connected old growth–dependent). Constraint (3.10) indicates that for every square kilometer of connected old growth, .426 species 2 animals result. The probability of connectivity (PR_i) is calculated in constraint (3.11) much like it was in the precise formulation in chapter 2. The primary distinction is the continuous (logistic) function between pr_{ij} and d_{ij} in constraint (3.12). Constraint (3.13) indicates that for every kilometer of edge (perimeter), 1.26 species 3 animals result.

Constraints (3.14)–(3.19) determine the population of species 4, depending on which constraints are limiting for each patch. Constraint (3.14) indicates that for every square kilometer of old growth in each patch, 41.2 animals result. Constraint (3.15) could also be the limiting factor for any given patch. It indicates that 27.4 animals result for every square kilometer of forageable open area within .366 km of each old growth circle. Constraint (3.16) calculates the area of this annulus of forage around each old growth circle. If an old growth circle is close to a border of the management area, part of its annulus of forage may fall outside the management area. Because the land outside the management area is assumed to be grassland, it will be included in the forage area calculation.

As currently formulated, the forage areas are allowed to overlap each other or an old growth circle, and this is not accounted for; the overlap is double-counted. For our purposes here, this inaccuracy does not seem to be terribly important, and is ignored (it is treated in the model in Hof and Joyce, 1992).

In constraint (3.14), the area variables are A_i' instead of the A_i. The A_i' are calculated in constraints (3.17) and (3.18) such that each $A_i' = A_i$ only if the r_i is greater than .86 km. If r_i is less than .86, then the associated A_i' is calculated to be 0. This reflects the threshold effect that each old growth patch must be a certain minimum size before any species 4 habitat results. Constraint (3.19) sums the individual species 4 patch populations into the S_4 population variable.

Results

All the solutions in this section were generated with the GRG2 algorithm, as previously indicated in chapter 1. Multiple starts were used, but global optima are not ensured. Also, in many of these solutions, numeric difficulties prevent absolute confidence in the precision of the results. The solution in figure 3.1 (and table 3.1) was generated with the formulation as specified by equations (3.1)–(3.19) and an objective function with the following weights:

$V_1 = 8$
$V_2 = 1$
$V_3 = 10$
$V_4 = 10$

The figures show the old growth circles retained. This solution uses all four potential circles to maximize the edge and species 4 forage area, given the optimal balance between area harvested and area left in old growth. All four circles are large enough to support species 4. The circles are located very close together so that the distance accessibility factor does not limit usable habitat (through constraints 3.12 and 3.16). This solution is quite consistent with the qualitative conclusions drawn by Diamond (1975) regarding desirable habitat patch layouts (e.g., that triangular or rectangular arrangements are more effective than linear ones). As an experiment, we imposed a solution on this model with four circles the exact size of those in figure 3.1, but arranged in a straight line (and just touching). As expected, the only species affected is species 2, which declined from the expected population of 9.0 in table 3.1 to an expected population of 7.3, because of the reduction in connectivity.

TABLE 3.1
Numerical Solutions for Figures 3.1–3.4

	Figure 3.1	Figure 3.2	Figure 3.3	Figure 3.4
S_1	872.60	469.60	348.90	—[a]
S_2	9.00	19.32	23.75	32.86
S_3	44.88	72.31	76.72	—[a]
S_4	403.33	621.68	656.76	—[a]
km² harvested	87.26	46.96	34.89	20.84
km² old growth	25.24	65.54	77.61	91.66
km² edge	35.62	57.38	60.86	55.63
Objective function	11,471.8	9481.1	8619.0	32.86

[a] Not valued.

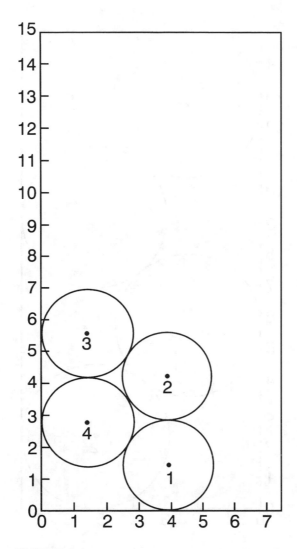

FIGURE 3.1

Example solution with $V_1 = 8$, $V_2 = 1$, $V_3 = 10$, and $V_4 = 10$.

The solution in figure 3.2 (and table 3.1) used the following objective function weights:

$V_1 = 5$
$V_2 = 10$

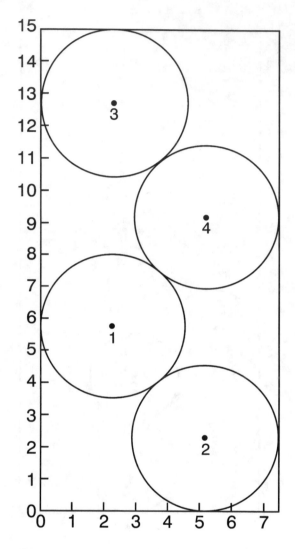

FIGURE 3.2
Example solution with $V_1 = 5$, $V_2 = 10$, $V_3 = 10$, and $V_4 = 10$.

$V_3 = 10$
$V_4 = 10$

With the lower species 1 value and higher species 2 weight, more old growth is retained, with a spatial configuration still very much like the

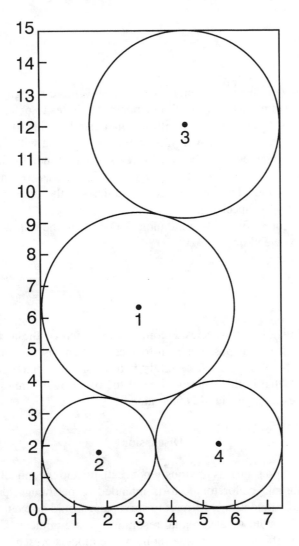

FIGURE 3.3
Example solution with $V_1 = 3$, $V_2 = 10$, $V_3 = 10$,
and $V_4 = 10$.

solution in figure 3.1. The elongation of the layout is the result of the rectangular planning area.

The solution in figure 3.3 (and table 3.1) used the following objective function weights:

$V_1 = 3$
$V_2 = 10$
$V_3 = 10$
$V_4 = 10$

With the lower species 1 weight, considerably more old growth is retained. The solution still uses all four potential circles to obtain the edge and species 4 forage area, but with unequal-sized circles so as to leave more of the area in old growth than would be possible with four equal-sized circles. The relative size of the circles is determined by the relative contributions of area, edge, and forage area factors, as well as the size and shape of the planning area as a whole. All the circles are large enough to support species 4.

The solution in figure 3.4 (and table 3.1) utilized the following objective function weights:

$V_1 = 0$
$V_2 = 1$
$V_3 = 0$
$V_4 = 0$

This maximization of species 2 commits as much of the area to connected old growth as possible with four circles. The asymmetrical solution connects the four circles slightly better than a symmetrical pattern would. Note that in this solution, one of the old growth circles is below the size threshold (species 4 is not valued).

Discussion

The example presented is obviously not realistic and the model formulations are initial, exploratory efforts intended to be thought-provoking. The nonlinear models are difficult to solve, and would be even more difficult to solve with greater spatial resolution (more circles). The principal solution difficulty is not time (none of the models presented required more than a few minutes to solve), but rather numeric precision. As the nonlinear models get larger and more complex, numeric imprecision causes the search algorithm to fail before solution times become a concern. Scaling of the formulation and tuning of the solution parameters definitely help, but are more difficult than with linear models and would vary with the solution algorithm used. However, the formulations presented do have the property that the number of choice variables increases as a linear function of the spatial resolution. In the example, only limited tradeoffs are demonstrated because there is really nothing in the

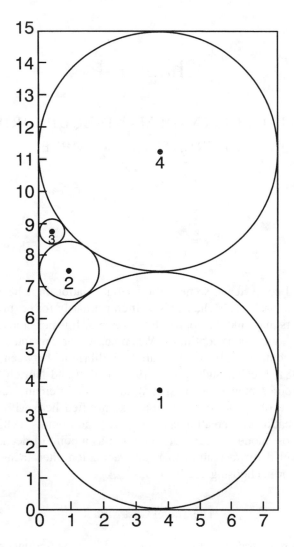

FIGURE 3.4
Example solution with $V_1 = 0$, $V_2 = 1$, $V_3 = 0$, and $V_4 = 0$.

formulation that penalizes proximity of the habitat patches. As a result, all the solutions are extremely well-connected. In chapter 6, we return to the geometric optimization approach to investigate the tradeoff between connectivity and spatial autocorrelation.

Chapter 4

SPATIAL SUPPLY–DEMAND EQUILIBRIUM: A RECREATION EXAMPLE

In managed ecosystems, some static spatial relationships have more of an economic basis. This chapter is written primarily for economists, but other readers may find the approach interesting. It is possible to skip this chapter without loss of continuity. When supply or demand is spatial in nature, economists analyze the "spatial equilibrium" between them (see Martin, 1981; McCarl and Spreen, 1980; Duloy and Norton, 1975). In natural resource management, spatial supply and demand factors may imply a type of spatial optimization, exemplified here with a case of strategic location of recreation sites (see Hof and Loomis, 1983). We use the travel cost model of recreation demand as a point of departure. Distances traveled by recreators to reach recreation sites then become a critical factor in choosing site locations.

The Travel Cost Model

The literature on travel cost models emphasizes the importance of taking into account the presence of an existing site for which the site under evaluation will substitute (perfectly). For example, Cicchetti et al. (1976), applying an approach suggested earlier by Burt and Brewer (1971), state,

Much of this chapter was adapted from J. G. Hof and J. B. Loomis, A recreation optimization model based on the travel cost method, *Western Journal of Agricultural Economics* 8(1) (1983): 76–85, with permission from the publisher, the Western Agricultural Economics Association.

Accordingly, any project or policy that results in a reduction in the travel time input t_i required in the production of services such as those provided by site i, may in turn be said to result in a reduction in price, P_i. One way in which this can be accomplished would be the development of a new site, close to the recreationist, which provides the same services. The strategy in any empirical application . . . is then to pick an existing site for which the new one can be assumed to (perfectly) substitute, and trace the effects of the price reduction through the system of derived demands [for the existing sites]. (p. 1262)

It should be noted that more recent literature has shown that only the own demand function for the single existing site is needed for a consumer surplus measure associated with a single price change (Just et al., 1982; Hof and King, 1982).

The Case of More Than One Proposed Site

In the case where there is more than one proposed site to be evaluated, the problem may be more complex. If the proposed sites could be regarded as independent new commodities that do not substitute for any other commodities, then they could theoretically be evaluated one at a time, in any order. On the other hand, if the proposed sites all substitute perfectly for an existing site, as discussed by Burt and Brewer, then the estimated value of any given proposed site would be affected by whether other (perfect substitute) proposed sites are developed. Therefore, an arbitrary order of evaluating proposed sites could seriously affect site development decisions. Also, the optimal solution might be partial development of some or all of the proposed sites rather than treating each site as an all-or-nothing project. What is needed is a method of simultaneously determining the levels and locations of recreation site development that will maximize net benefits. This would imply a cost-minimizing combination of travel from the potential origins and resource inputs for the proposed and existing sites.

The conventional means of calculating net benefits from a travel cost model is to deduct the cost of operating the new site, opportunity costs, and the amortized investment cost from the travel cost price-induced change in consumer surplus (see Burt and Brewer, 1971, p. 817). Another way of looking at this same net social value measurement is that travel costs, operating costs, opportunity costs, and investment costs are deducted from gross benefits measured by the first-stage demand function.[1] Looking at the problem in this way, the development of a set of

1. The first-stage demand curve is the direct regression of visits per capita against travel costs. The second-stage demand curve is derived by determining visitation to

new sites implies a change in the mix of inputs (sites and travel), which in turn leads to the price change and a change in net benefits.

With multiple proposed sites, recreators from different origins can potentially recreate at any of the proposed sites or at the existing site. Thus, on the supply side, the model we propose is somewhat like a transportation model because it seeks an efficient set of origin-to-destination "deliveries." On the demand side, prices are determined endogenously with demand functions (first stage) specified for different origins, so the model is somewhat like a spatial equilibrium model (see Martin, 1981). The approach represents both an optimization and a means of logically evaluating more than one proposed site simultaneously. Recreation visitor days (RVDs) at the different sites enter the solution according to efficiency (net benefit maximization) criteria, thus avoiding arbitrary ordering of the evaluation.

A Spatial Recreation Allocation Model

The basic choice variables in this model are the amounts of recreation, measured here in RVDs, to be consumed at the existing and at each proposed site by individuals from each origin.[2] A formulation for multiple proposed sites that all substitute (perfectly) for one existing site is as follows:

Maximize

$$\sum_{j=1}^{J} \int_{0}^{E_j} D_j(E_j)\,dE_j - \sum_{j=1}^{J}\sum_{i=1}^{I}(T_{ij} + M_{ij} + C_{ij})R_{ij}, \quad (4.1)$$

subject to

$$\sum_{i=1}^{I} R_{ij} = E_j \qquad \forall\, j, \quad (4.2)$$

$$\sum_{j=1}^{J} R_{ij} \le A_i \qquad \forall\, i, \quad (4.3)$$

---

be expected with a set of postulated increments in costs facing individuals at each origin. Burt and Brewer showed that for single sites, the sum of consumer surpluses from the first-stage demand curves (across origins) is equivalent to the area under the second-stage demand curve.

2. For the remainder of this discussion, the word *origin* is used to indicate either *zone* or *population center,* whichever is more appropriate for the given planning situation.

where

i indexes recreation sites,

j indexes recreationist travel origins (such as population centers),

E_j = number of RVDs per year from origin j,

D_j = first-stage demand function for the existing site and origin j,

R_{ij} = the number of RVDs per year at site i (including the existing site and all proposed sites) from origin j,

T_{ij} = the travel cost per R_{ij},

M_{ij} = the management cost per R_{ij},

C_{ij} = the opportunity cost per R_{ij},

A_i = the capacity in RVDs per year for site i,

J = the number of origins, and

I = the number of sites (existing and proposed).

Each origin's per capita first-stage demand function for the given existing site would typically be the same,[3] but each origin has different population levels and travel costs to the existing and proposed sites, so that aggregate demand functions vary across origins. Proper specification of the demand functions (D_j) includes all relevant cross price terms for imperfect substitute sites currently existing. Because these cross prices are held constant in the evaluation, however, they enter the linear programming objective function as constants. Thus, they can be ignored in this model. Because the model structure uses a linear programming solution procedure, the downward slope of each origin's demand curve is approximated in a piecewise fashion. The optimal solution to this problem assumes that recreators are cost-minimizers and that use is not allowed to exceed the site capacities.

The opportunity costs are generally the net value of commodities other than recreation that would be forgone because of recreational use of the land. Obviously, in some cases there may be no identifiable financial opportunity costs. It is expected that the site costs (M and C) for a given RVD at a given site are the same for all origins. The management costs, opportunity costs, and travel costs are straightforward calculations.

In the context of a regional planning model, several different types of recreation or several different market areas, distributed across the landscape, would typically be involved. The planner could build one model for each of these recreation types or market areas if net benefit estimation is all that is desired. If the model is to be used as a planning optimization tool, however, this may not be desirable. If a global agency

3. In fact, the per capita first-stage demand functions are typically regressed across origins in the travel cost model.

budget constraint or output target applies to all recreation types or market areas, then one model should be built for all of them. As long as the demand functions for these different types or market areas are independent, this is straightforward. All proposed sites must be placed in groups, with one existing site attached to each group as the site for which the proposed sites in that group substitute perfectly. Formulation (4.1)–(4.3) would be modified as follows:

Maximize

$$\sum_{j=1}^{J}\sum_{k=1}^{K}\int_{0}^{E_{jk}} D_{jk}(E_{jk})\,dE_{jk}$$

$$-\sum_{j=1}^{J}\sum_{k=1}^{K}\sum_{i=1}^{I_k}(T_{ijk} + M_{ijk} + C_{ijk})R_{ijk}, \quad (4.4)$$

subject to

$$\sum_{i=1}^{I_k} R_{ijk} = E_{jk} \qquad \forall\, j \qquad\qquad (4.5)$$

$$\forall\, k,$$

$$\sum_{j=1}^{J} R_{ijk} \le A_{ik} \qquad \forall\, i \qquad\qquad (4.6)$$

$$\forall\, k,$$

where
i indexes recreation sites within a given site group k,
j indexes recreationist travel origins,
k indexes site groups,
E_{jk} = number of RVDs per year from origin j visiting site group k,
D_{jk} = first-stage demand functions for site group k and origin j,
R_{ijk} = the number of RVDs per year from origin j to site i in group k,
T_{ijk} = the travel cost per R_{ijk},
M_{ijk} = the management cost per R_{ijk},
C_{ijk} = the opportunity cost per R_{ijk},
A_{ik} = the capacity in RVDs per year for site i in group k,
J = the number of origins,
I_k = the number of sites in site group k, and
K = the number of site groups.

This is the form of the model demonstrated in the example discussed later in this chapter. Equation (4.4) can be piecewise approximated with the following:

$$\sum_{j=1}^{J}\sum_{k=1}^{K}\sum_{h=1}^{H}V_{jkh}E_{jkh} - \sum_{j=1}^{J}\sum_{k=1}^{K}\sum_{i=1}^{I_k}(T_{ijk} + M_{ijk} + C_{ijk})R_{ijk},$$

$$\tag{4.7}$$

$$E_{jk} = \sum_{h=1}^{H}E_{jkh}, \tag{4.8}$$

$$E_{jkh} \leq \bar{L}_{jkh} \qquad \forall\, j \tag{4.9}$$
$$\forall\, k$$
$$\forall\, h,$$

where

h indexes approximation segments,

E_{jkh} = the hth segment of the demand function for origin j and site group k,

V_{jkh} = the midpoint demand function value for the hth segment of the demand function for origin j and site group k,

$\bar{L}_{jkh}$ = the segment length of the hth segment of the demand function for origin j and site group k, and

H = the number of segments.

In the case example, we used five segments to approximate each demand function, scaled to each origin and site group. It should be pointed out that if the demand functions for the existing sites included in the model are interrelated, then the problem becomes more complex. The objective function would include the line integral of the system of demand equations (see Just et al., 1982) for each origin

$$\sum_{j=1}^{J}\left\{\int_c \sum_{k=1}^{K}D_{jk}(E_{j1},\ldots,E_{jK})\,dE_{jk}\right\}$$

$$-\sum_{k=1}^{K}\sum_{j=1}^{J}\sum_{i=1}^{I_k}(T_{ijk} + M_{ijk} + C_{ijk})R_{ijk}, \quad (4.10)$$

in place of equation (4.4). Duloy and Norton (1975) present a means of incorporating such an objective function in a linear program for the case of symmetrical cross-partial demand derivatives. With ordinary Marshallian demand functions, the cross-partial demand derivatives are generally asymmetrical, and the line integral is generally path-dependent. Martin (1981) discusses an appealing approach to handling this sort of problem in a quadratic programming structure (see also McCarl and Spreen, 1980).

An Example

Here we discuss a simple demonstration model for part of the Colorado National Wilderness Preservation System from Hof and Loomis (1983). Obviously, this example application should not be given policy interpretation with regard to the suitability of specific wilderness areas. It is presented here solely to demonstrate the modeling approach and the types of data needed.

The Problem

Three existing wilderness areas (Eagles Nest, Rawah, and Weminuche) are taken as the sites for which six proposed sites might substitute. Eagles Nest is closest to Denver (70 miles west of Denver), Rawah is located 120 miles northwest of Denver, and Weminuche is in the southwest portion of Colorado (see figure 4.1). The proposed sites are the Further Planning Areas designated by the Roadless Area Evaluation Study II (RARE II, USDA Forest Service). It is assumed that the St. Louis Peak, Lost Creek, and Williams Fork Further Planning Areas will perfectly substitute for the Eagles Nest Wilderness Area; the Service Creek and Davis Peak Further Planning Areas will perfectly substitute for the Rawah Wilderness Area; and the Cannibal Plateau Further Planning Area will perfectly substitute for the Weminuche Wilderness Area. This application of the model could be viewed as a case where either three different types of wilderness are being provided, or three groups of wilderness sites are spatially removed such that they serve different markets.

Naturally, some of the proposed sites have some current use, even though they are not currently designated wilderness. For the purposes of this demonstrative example, this fact is ignored. The model obviously applies best to situations where proposed sites have no current use, such as a case where no access to the proposed sites is available without wilderness designation. For this simple example, only nearby origins (within roughly 300 miles) are included in the linear program (see figure 4.1). The Rawah site group origins are

Denver
Fort Morgan
Pueblo

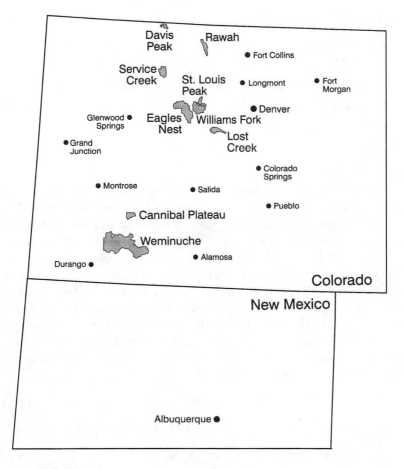

FIGURE 4.1
Existing sites, proposed sites, and demand origins for the case example.

Montrose
Longmont
Fort Collins

The Weminuche site group origins are

Salida
Denver
Grand Junction
Glenwood Springs
Fort Collins
Alamosa

Durango
Montrose
Colorado Springs
Albuquerque

The Eagles Nest site group origins are

Albuquerque
Denver
Colorado Springs
Grand Junction
Glenwood Springs

The travel cost demand curves were estimated from all relevant origin–destination data collected over several summers by the USDA Forest Service.[4] The weighted aggregate (zonal) travel cost method (see Bowes and Loomis, 1980) was used instead of the individual observation approach because the data did not track a particular individual's visitations per year. The general form of this simple travel cost demand function was visits per 1000 population as a function of the total round-trip travel costs. For more information on the demand function estimation, see Hof and Loomis (1983). The demand functions should be interpreted only as examples, not as reliable benefit estimators. The resulting per capita (actually per 1000 persons) demand equations for Weminuche, Eagles Nest, and Rawah, respectively, were

$$V/\text{pop} = .0424 - .0394 \ln \text{TC} \qquad (4.11)$$
$$(3.084) \qquad \bar{R}^2 = .335,$$

$$V/\text{pop} = .0345 - .0335 \ln \text{TC} \qquad (4.12)$$
$$(3.699) \qquad \bar{R}^2 = .564,$$

$$V/\text{pop} = .3000 - .0296 \ln \text{TC} \qquad (4.13)$$
$$(2.177) \qquad \bar{R}^2 = .290,$$

where
V/pop = visits per 1000 population and
TC = total travel cost in hundreds of dollars.

4. ZIP code and use data came from surveys directed by B. L. Driver and Perry J. Brown, funded by the USDA Forest Service, Rocky Mountain Forest and Range Experiment Station, and supported by Colorado State University McIntire–Stennis funds. The surveys were based on representative samples of summer-season users of the wilderness areas.

For use in this example, these equations were scaled to account for the sampling density of overall annual use. Visits were converted to RVDs at the rate of three RVDs per visit (USDA Forest Service, Rocky Mountain Region, unpublished data).

Site management costs of $1.75 per RVD (Walsh et al., 1981) were used. The opportunity costs and area capacities are given in table 4.1. Site capacity was set at one RVD per acre per year (Walsh et al.). The opportunity costs were constructed from the USDA Forest Service Development Opportunity Rating System (DORS) data gathered during the second Roadless Area Review and Evaluation (RARE II). DORS represents the net present value of commodity outputs (such as minerals and timber) given up in preserving wilderness. Because of the marginal nature of commodity output profitability due to remoteness, only two of the proposed sites had positive opportunity costs according to DORS. The annualized opportunity costs per acre were converted to costs per RVD using per acre recreation capacity. The cost assumptions are rough, but they serve for purposes of demonstration. Only variable costs were included in the linear program. For this wilderness example, investment costs were assumed to be 0.

Results

Table 4.2 presents the unconstrained efficiency solution for the example application. The results provide a foundation for comparison with allocations based on criteria other than efficiency. The solution in table 4.2 is intuitively appealing. For example, Lost Creek is small, but close to Denver and Colorado Springs (and other origins) and thus enters solu-

TABLE 4.1
Capacities and Opportunity Costs for Existing and Proposed Wilderness Areas

Area	Capacity (RVDs per year)	Opportunity Costs ($ per RVD)
Eagles Nest	130,915	0
St. Louis Peak	12,800	0
Williams Fork	74,700	0
Lost Creek	23,000	0
Rawah	27,464	0
Service Creek	39,860	.66
Davis Peak	11,532	0
Weminuche	401,400	0
Cannibal Plateau	31,990	.93

TABLE 4.2
Unconstrained Maximization of Net Benefits

Total benefits	$17,552,000
Travel costs	4,579,000
Management costs	425,000
Opportunity costs	29,750
Total costs	5,033,750
Net Benefits	$12,518,250

Area	RVDs	% of Capacity
Eagles Nest	37,844	29
St. Louis Peak	0	0
Williams Fork	21,565	29
Lost Creek	23,000	100
Rawah	12,319	45
Service Creek	0	0
Davis Peak	140	1
Weminuche	116,014	29
Cannibal Plateau	31,990	100
Total RVDs	242,872	

tion at capacity. Similarly, Cannibal Plateau is well located (though small) for the southern population centers in Colorado. The net benefits in table 4.2 are the optimal total net benefits for all existing and proposed sites. Thus, the current net benefits of existing sites should be deducted from the net benefits in table 4.2 if valuation of the new sites (in solution) is desired.

Table 4.3 presents a solution where a budget constraint ($250,000) is imposed on site management, resulting in reduced expansion of the wilderness system. In this example, the budget reduction from the optimal $425,000 to $250,000 results in a loss of net benefits on the order of $1.3 million.

Table 4.4 presents two solutions where target levels were imposed on the total number of RVDs produced by the system of sites (without any budget constraints). This has become common practice in public land management analysis. Table 4.4 shows that, in this example, the imposition of targets significantly affects the optimal solution. With a target of 500,000 RVDs (as compared to an optimal 242,872 RVDs), management costs are about twice those in the efficient solution, and the net benefit is reduced by about $9000. With a target of 750,000 RVDs, management costs are tripled, and a loss of about $1 million in net benefits is in-

TABLE 4.3
Run with $250,000 Budget Constraint on Management Costs

Total benefits	$14,043,000	
Travel costs	2,616,100	
Management costs	250,000	
Opportunity costs	29,750	
Total costs	2,895,850	
Net benefits	$11,147,150	

Area	RVDs	% of Capacity
Eagles Nest	7,749	6
St. Louis Peak	0	0
Williams Fork	10,783	14
Lost Creek	23,000	100
Rawah	6,500	24
Service Creek	0	0
Davis Peak	93	. 1
Weminuche	62,688	16
Cannibal Plateau	31,990	100
Total RVDs	142,803	

TABLE 4.4
Analysis of Target Levels

	Target = 500,000 RVDs	Target = 750,000 RVDs
Total benefits	$18,380,000	$19,599,000
Travel costs	4,966,200	6,692,400
Management costs	875,000	1,312,500
Opportunity costs	29,750	56,100
Total costs	5,870,950	8,061,000
Net benefits	$12,509,050	$11,538,000

Area	RVDs (% capacity)	RVDs (% capacity)
Eagles Nest	37,844 (29)	130,915 (100)
St. Louis Peak	0 (0)	12,800 (100)
Williams Fork	21,565 (29)	74,770 (100)
Lost Creek	23,000 (100)	23,000 (100)
Rawah	27,464 (100)	27,464 (100)
Service Creek	0 (0)	39,860 (100)
Davis Peak	140 (1)	7,801 (68)
Weminuche	357,997 (89)	401,400 (100)
Cannibal Plateau	31,990 (100)	31,990 (100)
Total RVDs	500,000	750,000

curred. If the benefit measures are reliable, then imposition of target levels is illogical and can affect optimal solutions significantly. In the common situation where targets are imposed because of a lack of confidence in benefit measures, it must be recognized that benefit measurement has not been avoided; it has been done implicitly by setting the target levels. Setting targets in a tenable manner may be just as difficult as measuring benefits in a tenable manner.

Discussion

This chapter started with a conceptual view of travel cost models such that a new site's net benefits are created by the implicit recreation price decreases associated with that new site's development. In that context, if a number of proposed sites are included in a regional planning problem, then optimizing across these proposed sites as they substitute (perfectly) for an existing site is a very natural extension of the traditional travel cost model. Because of the origin–destination structure of the problem with downward-sloping demand functions, its formulation has a definite spatial character, but one that is quite different from the others discussed in this book.

PART II

SPATIAL AUTOCORRELATION

Part I discussed basic, static spatial relationships. Another type of spatial relationship is of a stochastic nature. We have come to expect managed ecosystems to exhibit considerable random (some also say chaotic) behavior. In addition, it is widely recognized that this randomness often includes a significant spatial component, or spatial autocorrelation (see chapters 5 and 6 for specific topical references). Spatial autocorrelation, where the covariance in ecosystem response is systematically ordered across locations, is to be expected because many sources of disturbance are spatially defined. Proximity implies shared vulnerability to weather, fire, insect and disease outbreaks, and so forth. Proximity also implies shared properties that are defined spatially across the landscape including soil properties, geography, elevation, and location in large-scale ecological biomes. It is hard to visualize a source for randomness that does not at least have the potential to be spatially covariant. Highly deterministic site characteristics such as aspect or elevation are often accounted for through stratification of variables in a model, but similarities in highly stochastic effects are not so easily captured.

The most common (and probably the most powerful) way to account for randomness in response coefficients in optimization is the chance constraint. There are actually a number of different types of chance con-

Some of this introduction was adapted from Hof, I G., B. M. Kent, and J. B. Pickens, Chance constraints and chance maximization with random yield coeeficients in renewable resource optimization, *Forest Science* 38(2) (1992): 305–323, with permission from the publisher, the Society of American Foresters.

straints and ways to use them, and we briefly discuss six of them here (see Hof et al., 1992). All these approaches directly include the variance-covariance matrix between coefficients within a given row, or constraint, in an optimization model, so that spatial autocorrelation (as well as other correlations) between those coefficients can be accounted for. At this time, covariances between rows cannot be analyzed in chance constraints. Chapters 5–7 demonstrate these approaches with specific regard to handling spatial autocorrelation, and clarify the limitations presented by our inability to account for between-row covariances.

Chance-Constrained Programming

Individual Chance Constraints

Charnes and Cooper (1963) provide what has become a classic analysis of chance constraints in linear programming. Their analysis focuses on problems where some or all of the right-hand sides are random variables. Charnes and Cooper's approach to this problem is to derive linear, deterministic formulations that are equivalent to the probabilistic problems. Their approach applies to the case where one wishes to constrain the model such that *each* right-hand side (input or output) is obtained with a prespecified probability.

Van de Panne and Popp (1963) extended the Charnes and Cooper analysis to the case of a random A-matrix (the matrix of decision variable coefficients for all constraints in the model). They start with a constraint of the form

$$\sum_{j=1}^{n} a_{ij} x_j \geq b_i \quad \text{for a given } i. \tag{II.1}$$

Van de Panne and Popp assume that the a_{ij} are stochastically independent but this assumption can be relaxed by applying the covariance matrix as in Miller and Wagner (1965). Assuming that the a_{ij} have distributions with means α_{ij} and variances/covariances σ_{ijh}^2,[1] a chance constraint that requires a γ_i probability of meeting the right-hand side b_i can be written as

---

1. We use σ_{ijh}^2 here as a symbol for $\rho_{ijh}\sigma_{ij}\sigma_{ih}$, the variance/covariance between a_{ij} and a_{ih}, where ρ_{ijh} is the correlation coefficient of a_{ij} and a_{ih}, with standard deviations σ_{ij} and σ_{ih}, respectively. Variance is just the special case where $j = h$ (and $\rho_{ijh} = 1$).

$$\sum_{j=1}^{n} \alpha_{ij}x_j + \delta_i \left(\sum_{j=1}^{n} \sum_{h=1}^{n} x_j x_h \sigma_{ijh}^2 \right)^{1/2} \geq b_i, \qquad \text{(II.2)}$$

where δ_i is the standard deviate (sometimes called the tabular value, such as the z value for normally distributed random variables) corresponding to the required probability, such that

$$\gamma_i = G(b_i^*) \equiv 1 - F(b_i^*), \qquad \text{(II.3)}$$

where $F(b_i^*)$ is the cumulative probability density function for the row total $b_i^* = \sum_j \alpha_{ij}x_j^*$ that results from a particular solution x_j^* for all j. The ith row total will have a mean of $\sum_j \alpha_{ij}x_j$ and a variance of $\sum_{j=1}^{n} \sum_{h=1}^{n} x_j x_h \sigma_{ijh}^2$, which leads directly to formulation (II.2). It should be clear from (II.2) that if a .5 probability of meeting b_i is desired when the probability density function $f(b_i^*)$ is symmetric about the mean, the spatial autocorrelations are not relevant (in fact, the entire variance-covariance matrix is irrelevant). With a .5 confidence level, $\delta_i = 0$ and only the mean levels of the a_{ij} (the α_{ij}) are needed. Another way of interpreting this point is that if means of the a_{ij} are inserted into the original linear program and variances/covariances are ignored, the solution should be interpreted as having a .5 probability of meeting each binding constraint individually. For convenience, we assume normally distributed a_{ij} throughout this section. This assumption results in normally distributed b_i^* row totals as well, because they are linear functions of the normally distributed a_{ij}. This allows us to use z values for the δ_i standard deviates.

With independence between rows, any number of these types of constraints can be included in the overall mathematical program. This mathematical program is nonlinear, but Van de Panne and Popp show that (for normally distributed a_{ij}) it is convex if all $\gamma_i \geq 1/2$. In this case, only a single, global optimum exists. The same between-row independence is necessary for all the formulations that follow. Relaxation of this assumption is, at this time, analytically intractable. In renewable resource optimization problems with multiple time periods, coefficients in different rows may be interrelated because one time period's status or yield is indirectly correlated with another time period's status or yield. The impact of ignoring between-row covariances when they are, in fact, nonzero is difficult to predict. This topic is discussed further in chapter 5.

Joint Probability Chance Constraint

Miller and Wagner (1965) extend the Van de Panne and Popp approach to the case where it is desired to constrain the joint probability of meeting a set of right-hand sides to be at least some prespecified constant β, rather than to limit each constraint to meet its right-hand side with a certain probability (see also Jagannathan, 1974; Brown and Rutemiller, 1977, Balintfy, 1970). That is, a joint chance constraint that replaces q of the original constraints, of the form

$$\prod_{i=1}^{q} \Pr\left(\sum_{j=1}^{n} a_{ij}x_j \geq b_i\right) \geq \beta, \tag{II.4}$$

is included. Miller and Wagner formulated this as

$$\sum_{j=1}^{n} \alpha_{ij}x_j + y_i\left(\sum_{j=1}^{n}\sum_{h=1}^{n} x_j x_h \sigma_{ijh}^2\right)^{1/2} \geq b_i \qquad i = 1,\ldots,q, \tag{II.5}$$

$$\prod_{i=1}^{q} G_i(y_i) \geq \beta, \tag{II.6}$$

where each y_i (analogous to the δ_i used in the Individual Chance Constraints formulation) is a standard normal deviate, and is now a choice variable. The $G_i(y_i)$ are defined by

$$G_i(y_i) \equiv 1 - F(y_i) \qquad i = 1,\ldots,q, \tag{II.7}$$

and $F(y_i)$ is a standard normal cumulative density function (CDF). This formulation allows the optimization process to allocate uncertainty across the q constraints (by treating the y_i as choice variables) while meeting the joint chance constraint (II.6). Unfortunately, this does not necessarily lead to a convex program (regardless of the value of β), so local optima are possible. Because the y_i are choice variables in this formulation, it is necessary to approximate the closed form of the cumulative density function (F). Two of many possible approximations are described at the end of this introduction.

Total Probability Chance Constraint

An alternative to the Joint Probability Chance Constraint approach applies when it is desired to constrain the expected value of the number of right-hand sides that are met.[2] This is equivalent to constraining the total

2. Sengupta (1972:71) suggests an arithmetic mean measure of reliability across multiple uncertain components. Also, Fox et al. (1966) provide a general discussion of decision rules under chance-constrained programming.

of the individual constraint probabilities to be at least some prespecified constant (say, Φ):

$$\sum_{i=1}^{q} \Pr\left(\sum_{j=1}^{n} a_{ij}x_j \geq b_i\right) \geq \Phi. \tag{II.8}$$

Modifying the Joint Probability Chance Constraint approach accordingly yields

$$\sum_{j=1}^{n} \alpha_{ij}x_j + y_i\left(\sum_{j=1}^{n}\sum_{h=1}^{n} x_j x_h \sigma_{ijh}^2\right)^{1/2} \geq b_i \qquad i = 1,\ldots,q, \tag{II.9}$$

$$\sum_{i=1}^{q} G_i(y_i) \geq \Phi, \tag{II.10}$$

where all variables are defined as before. Again, there is no assurance that this is a convex program, so local optima are possible and it is necessary to approximate the cumulative density function (F).

Chance-Maximizing Programs

In renewable resource management problems, it may be desired to maximize the chance of meeting targets or input limits, given a random A-matrix. With a single function, one could simply convert (II.2) into an equality, and then either minimize δ_i subject to a fixed b_i, or maximize b_i subject to a given level of δ_i. For multiple functions, three approaches can be identified, each being a counterpart to one of the chance-constrained approaches previously outlined.[3]

MAXMIN Chance-Maximizing Programming

The chance-maximizing counterpart to the Individual Chance Constraints approach is to maximize the minimum probability of meeting a right-hand side across all $i = 1, \ldots, q$ constraints (see Sengupta, 1972:73):

Minimize λ
subject to

$$\lambda \geq y_i \qquad i = 1,\ldots,q, \tag{II.11}$$

---

3. Sengupta (1972) suggested including joint probabilities in objective functions in terms of reliability measures.

$$\sum_{j=1}^{n} \alpha_{ij} x_j + y_i \left(\sum_{j=1}^{n} \sum_{h=1}^{n} x_j x_h \sigma_{ijh}^2 \right)^{1/2} \geq b_i, \tag{II.12}$$

where λ is the largest of the y_i. By minimizing λ, the minimum probability is maximized. In contrast to the Individual Chance Constraints approach, the y_i are now treated as choice variables, and can be transformed into probabilities by

$$\gamma_i = G_i(y_i) \equiv 1 - F(y_i), \tag{II.13}$$

typically with a standard normal table. As with the Individual Chance Constraints formulation, this is a convex program if all $\gamma_i \geq 1/2$. Thus, if all γ_i solve with a value $\geq 1/2$, that solution is a global optimum. It is conceivable that a program with this formulation might solve with a local optimum where one or more of the γ_i are $\leq 1/2$ even though the global optimum involves all $\gamma_i \geq 1/2$.

It should be noted that the Individual Chance Constraints and MAXMIN Chance Maximizing approaches could be formulated in a Miller and Wagner-type formulation with an approximated CDF. Although this requires the approximation, it provides a formulation that is somewhat easier to interpret (because the CDFs are explicitly specified). The Individual Chance Constraint Miller and Wagner-type formulation for the ith constraint is

$$\sum_{j=1}^{n} \alpha_{ij} x_j + y_i \left(\sum_{j=1}^{n} \sum_{h=1}^{n} x_j x_h \sigma_{ijh}^2 \right)^{1/2} \geq b_i, \tag{II.14}$$

$$G_i(y_i) \geq \gamma_i. \tag{II.15}$$

The MAXMIN Chance Maximizing Miller and Wagner-type formulation is as follows:

Maximize ρ

subject to

$$\rho \leq G_i(y_i) \qquad i = 1, \ldots, q, \tag{II.16}$$

$$\sum_{j=1}^{n} \alpha_{ij} x_j + y_i \left(\sum_{j=1}^{n} \sum_{h=1}^{n} x_j x_h \sigma_{ijh}^2 \right)^{1/2} \geq b_i, \tag{II.17}$$

where ρ, the smallest of the γ_i, is being maximized.

Joint Probability Chance-Maximizing Programming

The chance-maximizing counterpart to the Joint Probability Chance Constraint approach is to maximize the joint probability that all the q right-hand sides are met:

Maximize

$$\prod_{i=1}^{q} \Pr\left(\sum_{j=1}^{n} a_{ij}x_j \geq b_i\right).$$ (II.18)

Modifying the Joint Probability Chance Constraint formulation accordingly yields the following:

Maximize

$$\prod_{i=1}^{q} G_i(y_i),$$ (II.19)

subject to

$$\sum_{j=1}^{n} \alpha_{ij}x_j + y_i \left(\sum_{j=1}^{n}\sum_{h=1}^{n} x_j x_h \sigma_{ijh}^2\right)^{1/2} \geq b_i \qquad i = 1, \ldots, q.$$ (II.20)

As with the Joint Probability Chance Constraint approach, the $G_i(y_i) = 1 - F(y_i)$ and the closed form of this cumulative density function (F) must be approximated. Again, there is no assurance that this is a convex program, so local optima are possible.

Total Probability Chance-Maximizing Programming

The chance-maximizing counterpart to the Total Probability Chance Constraint approach is to maximize the total probability that all the q right-hand sides are met:

Maximize

$$\sum_{i=1}^{q} \Pr\left(\sum_{j=1}^{n} a_{ij}x_j \geq b_i\right).$$ (II.21)

Modifying the Joint Probability Chance Constraint formulation accordingly yields the following:

Maximize

$$\sum_{i=1}^{q} G_i(y_i),$$ (II.22)

subject to

$$\sum_{j=1}^{n} \alpha_{ij} x_j + y_i \left(\sum_{j=1}^{n} \sum_{h=1}^{n} x_j x_h \sigma_{ijh}^2 \right)^{1/2} \geq b_i \qquad i = 1, \ldots, q.$$

(II.23)

As with the Total Probability Chance Constraint approach, the $G_i(y_i) = 1 - F(y_i)$ and the closed form of this cumulative density function (F) must be approximated. There is no assurance that this is a convex program, so local optima are possible.

Approximation of the CDF

In the Joint Probability Chance Constraint, Total Probability Chance Constraint, Joint Probability Chance-Maximizing, and Total Probability Chance-Maximizing approaches, it is necessary to approximate the closed form of the standard normal cumulative density function (F). Two workable approaches are a logistic approximation and a polynomial approximation. A simple logistic approximation is

$$G_i(y_i) \equiv 1 - F(y_i) \approx 1 - \frac{1}{1 + e^{-y_i/.5}}.$$

(II.24)

This approach has a probability (F) equal to .5 at "mean" 0, and 96.4 percent of the probability is contained in the interval plus or minus four times the scale parameter (.5), thus approximating a normal curve having a standard deviation of 1. It should be noted that other logistic approximations could be obtained by varying the scale parameter. The specific scaling used determines where the approximation will be most precise. An example polynomial approximation, from Abramowitz and Stegun (1965:932), is

$$G_i(y_i) \equiv 1 - F(y_i) \approx f(y_i)(a_1 t + a_2 t^2 + a_3 t^3),$$

(II.25)

where
$t = 1/(1 + p y_i),$
$p = .33267,$
$a_1 = .4361836,$
$a_2 = -.1201676,$
$a_3 = .937298,$ and
$f(y_i) = (1/\sqrt{2\pi}) e^{-y_i^2/2}.$

There are other polynomial approximations that are more complex (and more accurate) and less complex (and less accurate). We selected this approximation as an example because it avoids very large or small numbers that might cause numeric problems in a nonlinear program, is not too complicated, and still has excellent accuracy (within $\pm 1 \times 10^{-5}$). A polynomial approximation such as (II.25) makes the nonlinear program larger as well as requiring more calculations, so one would expect the logistic approach to be easier to work with and solve. This might be especially important in larger problems.

In chapter 5, we demonstrate the use of an individual chance constraint to account for spatial autocorrelation in timber yields in a static model. In chapter 6, we demonstrate a simple chance maximization to account for wildlife population/habitat autocorrelation, taken in combination with connectivity considerations. The extension to multiple chance constraints and the other, more complicated approaches is straightforward, but with possible problems in solving the nonlinear programs. Chapter 7 treats the case of large models with multiple chance constraints where nonlinear methods are not practical.

Chapter 5

A CELLULAR TIMBER MODEL
WITH SPATIAL AUTOCORRELATION

Recent work has demonstrated the importance of accounting for risk and uncertainty in optimizing forest management decisions (Pickens and Dress, 1988; Weintraub and Vera, 1991; Hof et al., 1992). These analyses center on random technical coefficients (the yield predictions under alternative management prescriptions) in natural resource linear programming models. At the same time, a considerable amount of literature is appearing (see Matérn 1986, 1993; Reed and Burkhart, 1985; Czaplewski et al., 1994) that suggests that not only are timber yield and other ecosystem response coefficients random, they can be spatially autocorrelated. We focus in this chapter on a rather simple spatial autocorrelation structure for timber yields, such that the yield coefficients have covariances that are defined by proximity. The purpose of this chapter is to model and analyze the impact of spatial yield covariances on the optimal layout of management actions in a chance-constrained formulation. We do so with a simple example problem (from Hof et al., 1996).

This chapter was adapted largely from J. Hof, M. Bevers, and J. Pickens, Chance-constrained optimization with spatially autocorrelated forest yields, *Forest Science* 42(1) (1996): 118–123, with permission from the publisher, the Society of American Foresters.

An Example

The Problem

Suppose we must lay out timber harvests on an area as mapped in figure 5.1a, which includes a river and two access roads (similar to the example in chapter 2). Let us once again discretize the land base, as in figure 5.1b, into 25 cells and approximate locations for the river and roads. Assume that we are uncertain about the actual timber volumes in each cell, but we have reliable estimates of the mean yield and variance. The area is all the same age, with a mean yield of 120 cunits (units of timber volume) per cell and a coefficient of variation of 0.5, so that the standard deviation of yield per cell is 60 (variance = 3600).

Next we assume that the yield covariances σ_{ij}^2 are defined between cell i and cell j as

$$\sigma_{ij}^2 = \rho_{ij}(60^2), \tag{5.1}$$

where ρ_{ij}, the autocorrelation coefficient between the yield coefficient in cell i and the yield coefficient in cell j, is a function of the distance between cells i and j:

$\rho_{ij} = 1$ if $i = j$ (σ_{ij}^2 is the variance),
$\rho_{ij} = .7$ if i and j are adjacent
$\rho_{ij} = .175$ if i and j are one cell removed
$\rho_{ij} = 0$ if i and j are farther apart

This approximates a function

$$\rho_{ij} = \frac{.7}{d_{ij}^2}, \tag{5.2}$$

where d_{ij} is the distance between cell i and cell j centroids, in units of cell-side-length. Let us also assume that the river and roads create obstacles such that the covariances across these obstacles are 0 (because, for example, they prevent spread of fire or pathogens or they mark boundaries of different, independent tree species).

This is a simple specification of spatial autocorrelation, but is not unreasonable. For example, Weintraub and Vera (1991:784) state, "To define correlations we assumed a geographical pattern for the different timber stands and assigned correlation coefficients based on the distances between them." The basic reason for spatial autocorrelation is that distance often determines the commonality of random influences on the

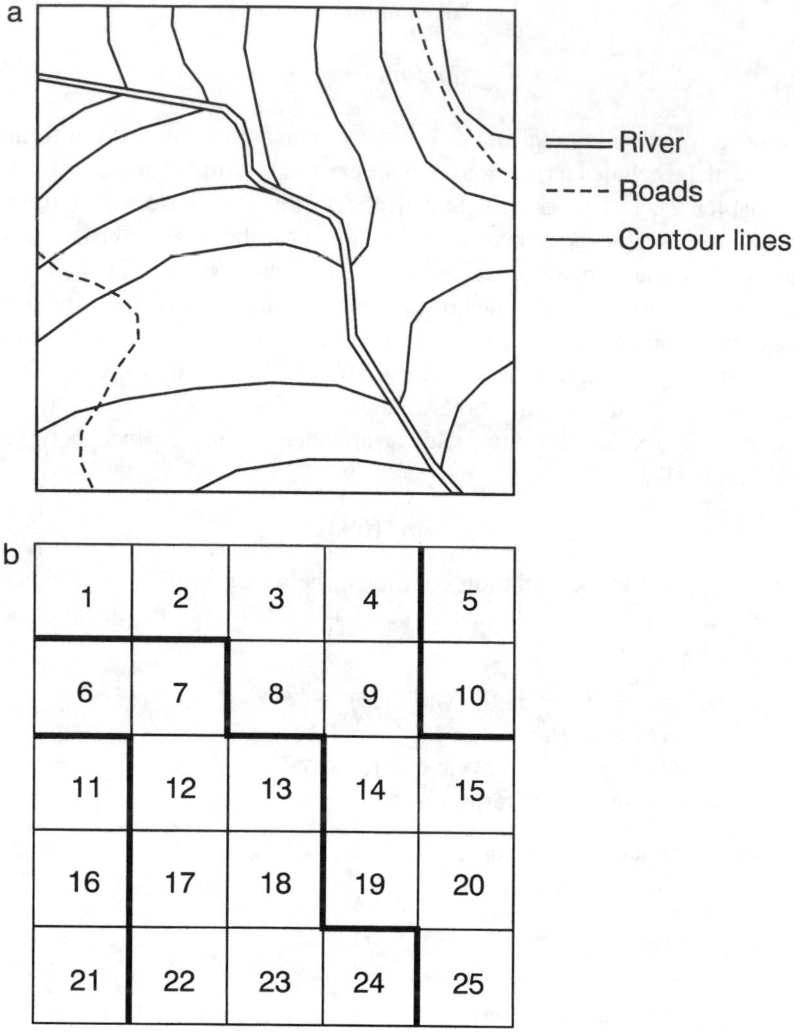

FIGURE 5.1
A mapped planning area (a) and its 25-cell digitized equivalent (b).

yield predictions (weather, fire, pathogen infestation, etc.). Covariances over time are briefly discussed, but again, the focus of this chapter is the case of spatially defined covariances. Therefore a static model is used, applying to the spatial layout of harvests in any given time period with the covariance structure as defined.

Given the problem as specified, suppose we wish to minimize area-based harvest cost (with known cost coefficients), such that we are 95 percent confident of achieving a given timber output level (for the given time period). As shown in the introduction to part II, this problem can be formulated as the following nonlinear program (see Hof et al., 1992; Van de Panne and Popp, 1963):

Minimize

$$\sum_{i=1}^{25} C_i X_i, \tag{5.3}$$

subject to

$$\sum_{i=1}^{25} a_i X_i + \delta \left(\sum_{i=1}^{25} \sum_{j=1}^{25} X_i X_j \sigma_{ij}^2 \right)^{1/2} \geq b, \tag{5.4}$$

$$0 \leq X_i \leq 1 \quad \forall \, i,$$

where

X_i = a choice variable for each cell that represents the portion of that cell harvested,

C_i = the deterministic cost of harvesting the ith cell (assumed completely linear and divisible),

a_i = the mean yield for the ith cell,

δ = the z value or standard normal deviate for the given confidence level ($\delta = -1.65$ for 95 percent confidence),

σ_{ij}^2 = the covariance between cell i and j as defined previously, and

b = the output (timber) target for which we desire 95 percent confidence.

This approach thus optimizes the layout of harvests so as to achieve 95 percent confidence of meeting the timber target at minimum cost. The solution simultaneously finds the mean level of total output and amount of total output variance that minimizes (5.3) while meeting (5.4). The chance constraint (5.4) emanates from the linear summation of harvests, such that the *total* output has a mean of $\sum_i a_i X_i$ and a variance of $\sum_i \sum_j X_i X_j \sigma_{ij}^2$ (the X_i are constant in any given solution, but the a_i are randomly distributed).

How might we expect the spatially defined covariances to affect the spatial layout of harvests? Obviously constraint (5.4) is "easier" to meet (and can often be met at lower cost) if the total output variance can be reduced. It has long been known that the variance of an investment portfolio can be reduced through diversification of that portfolio, as long as the covariances between portfolio components are small relative to their variances (see Markowitz, 1959). Applying this logic to a land allocation problem such as ours would imply diversifying the harvest in any cutting period across cells. As a simple example, assume a given harvest area that can be allocated to $i = 1, \ldots, n$ cells with mean yields v_i, equal variances σ^2, and 0 covariances between the yields. Within the cell size limits, if γ_i is the proportion of required harvest area allocated to the ith cell ($\sum_i \gamma_i = 1$) and if the total yield is R, then

$$E(R) = E\left(\sum_{i=1}^{n} \gamma_i v_i\right), \tag{5.5}$$

$$\text{Var}(R) = \sum_{i=1}^{n} \gamma_i^2 \sigma^2, \tag{5.6}$$

so that Var(R) is smaller than σ^2 unless one $\gamma_i = 1$, in which case Var(R) $= \sigma^2$. This demonstrates the *potential* for reducing output–total variances through diversification when covariances between cells are small. In contrast, if the covariances between cells are large, little potential would be expected for reducing row-total variances through diversification. To demonstrate, let us again assume equal variances σ^2, but now assume that all covariances are σ^2 as well. Now,

$$\text{Var}(R) = \sum_{i=1}^{n}\sum_{j=1}^{n} \gamma_i \gamma_j \sigma^2, \tag{5.7}$$

so that Var(R) $= \sigma^2$ regardless of the γ_i allocation.[1]

Because the covariances in our problem diminish with distance, optimal spatial layouts tend to diversify any given harvest spatially. Not only is any given harvest dispersed across multiple cells, the cells chosen for any harvest tend to be dispersed widely across the landscape. Evaluation of the desirability of this outcome in terms of logging costs and logistics is beyond the scope of this chapter, but we explore trade-offs between autocorrelations and wildlife habitat connectivity considerations in chapter 6.

Before proceeding we should observe that the chance constraint (5.4) captures static yield covariances across space, but ignores yield covariances over time. In that sense, the formulation does not provide the choice variables or relationships to manage a particular source of uncer-

---

1. Var(R) $= \sigma^2 \sum_i \sum_j \gamma_i \gamma_j$, which is $\sigma^2 \sum_i \gamma_i$ because $\sum_j \gamma_j = 1$, which, in turn, is σ^2 because $\sum_i \gamma_i = 1$.

tainty (such as a pest or fire risk) over time through isolation of high-risk areas. The formulation applies only to a single harvesting action and the uncertain yields from that action. A dynamic analysis that manages a source of uncertainty through isolation with this type of formulation would be difficult because covariances between yield coefficients in different constraints (that apply to different time periods) would be inevitable, and no formulation for chance-constraining with between-row covariances is currently available (Miller and Wagner, 1965; Hof et al., 1992). Process-oriented dynamic models, as discussed in part III, may show more promise along these lines.

Results

We initially assumed uniform costs of $1000 per cell harvested (where $ is an arbitrary monetary unit). The model in (5.3) and (5.4) was then solved with a variety of values set for b. Nonlinear solver tolerances were set such that feasibility is assured within 1×10^{-5} for all solutions presented. With the confidence level set above 50 percent, (5.3) and (5.4) constitute a convex program (Van de Panne and Popp, 1963), so local optima are not a problem.

Figure 5.2a presents the spatial layout of harvests in solution with $b = 400$. It is clear that the spatial covariances cause the solution to disperse the harvest spatially. In order to arrive at an indication of the effects of this dispersal on expected timber output, we solved a corollary model:

Minimize

$$\hat{b}, \tag{5.8}$$

subject to

$$\sum_{i=1}^{25} X_i \geq \bar{A}, \tag{5.9}$$

$$\hat{b} = \sum_{i=1}^{25} a_i X_i + \delta \left(\sum_{i=1}^{25} \sum_{j=1}^{25} X_i X_j \sigma_{ij}^2 \right)^{1/2}, \tag{5.10}$$

$$0 \leq X_i \leq 1 \qquad \forall\, i,$$

where
$\bar{A}$ = the number of cells harvested in the previous solution.

This model basically finds the worst spatial allocation of the previous harvest, in terms of the 95 percent confidence level of expected timber

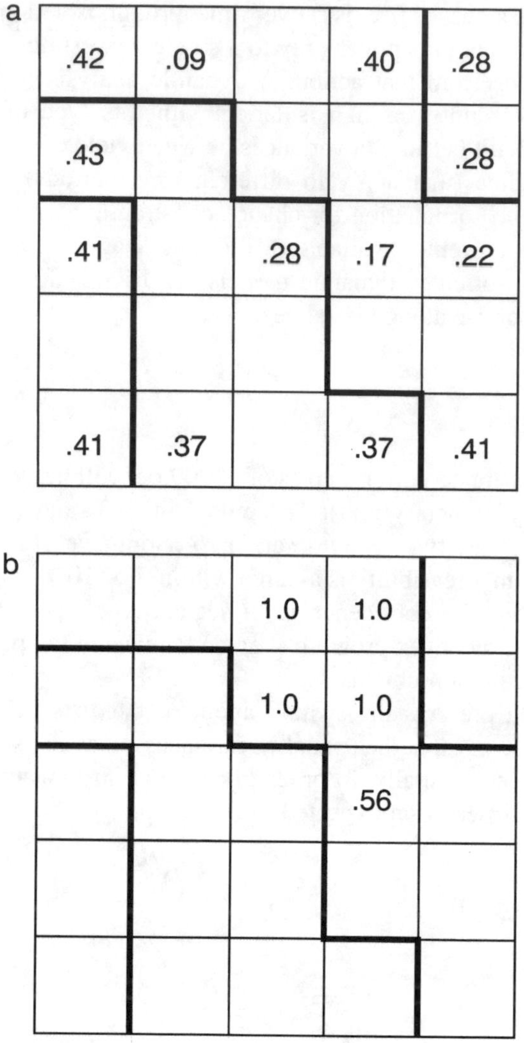

FIGURE 5.2
Optimal proportions of cells harvested for chance-constrained cost minimization (a) and worst-case harvest pattern (b), with uniform costs and $b = 400$ (blank indicates zero harvest).

output (b).[2] Notice that with homogeneous mean yields, (5.9) is also, in effect, a constraint on the mean total timber output. The spatial layout of harvest from solving this model is given in figure 5.2b. The harvest is now clumped together in cells 3, 4, 8, 9, and 14 (see figure 5.1b for cell numbers). The first row in table 5.1 gives the numerical results of these two runs. The spatial layout in figure 5.2a, obtained with (5.3) and (5.4), yields a minimum cost of $4,557 and a harvest of 4.557 cells. When (5.8), (5.9), and (5.10) were solved, the $\hat{b}$ obtained was 167.82. This indicates that if an analyst ignored the spatial layout of the harvest (everything else is constant across cells at this point), the timber output attain-

TABLE 5.1
Case Example Solutions

	Cost	Cells Harvested	$\hat{b}$ (minimized 95 percent confidence output)
Uniform Costs			
$b = 400$	4,557	4.557	167.82
$b = 800$	9,115	9.115	488.24
$b = 1200$	13,749	13.749	927.26
$b = 1600$	18,659	18.659	1,369.83
$b = 2000$	24,181	24.181	1,965.73
First Cost Modification			
$b = 400$	9,815	4.908	287.91
$b = 800$	19,885	9.943	757.76
$b = 1200$	32,488	14.775	1,170.59
$b = 1600$	52,957	19.066	1,494.51
$b = 2000$	77,318	24.182	1,968.82
Second Cost Modification			
$b = 400$	10,409	5.205	307.92
$b = 800$	21,608	10.602	744.72
$b = 1200$	34,811	15.983	1,187.26
$b = 1600$	52,459	19.720	1,479.04
$b = 2000$	75,128	24.251	1,973.42

2. We should note that this corollary model is not necessarily convex, so local optima are possible. We used multiple starts, and all results reported appear to be global optima. Because this model is used to determine a worst case, our results are conservative; if any of the optima reported are nonglobal, the implications are actually stronger than those discussed.

ment at 95 percent confidence might be underachieved by as much as 232.18 cunits. This result suggests that if spatial covariances between yields are significant, then it is potentially quite important to account for them in analyzing (optimizing) harvest decisions.

The second through fifth lines of table 5.1 present similar results for successively larger values of b. As one might expect, as b increases, the number of cells that must be harvested increases, and $\hat{b}$ converges on b because fewer and fewer spatial options remain available.

These results reflect the assumed homogeneity of the initial example. To examine a less homogeneous case, we redefined costs, by cell, two different ways as shown in table 5.2. The first cost modification assumes that harvested timber will be removed via the roads, so each cell's cost is a positive function of its distance from the nearest road. The second cost

TABLE 5.2
Cost Modifications

Cell	First Cost Modification	Second Cost Modification
1	4500	2000
2	5000	2000
3	4500	2500
4	2000	5000
5	2000	5500
6	2000	2000
7	2500	2000
8	4500	2000
9	2000	2500
10	2000	4500
11	2000	4500
12	2000	2500
13	4500	2000
14	2500	2000
15	2000	4500
16	2000	5000
17	2000	4500
18	4500	2000
19	5000	2000
20	4500	2500
21	2000	5500
22	2000	5000
23	4500	2500
24	5500	2000
25	5000	2000

modification assumes that harvested timber will be removed via the river, so each cell's cost is a positive function of its distance from the river. The first cost modification tends to cluster the harvested area around the two roads, whereas the second tends to cluster the harvested area along the river. This is traded off in the optimization against the tendency of the spatial covariances to generally disperse the harvest.

Figure 5.3 presents the layout of harvests for formulation (5.3)–(5.4) (figure 5.3a) and for formulation (5.8)–(5.10) (figure 5.3b), with $b = 400$ and the first cost modification. Also, the middle section of table 5.1 presents the ancillary numerical results. Because cost is no longer uniform across cells, an upper bound on cost was added to (5.8)–(5.10) in determining $\hat{b}$:

$$\sum_{i=1}^{25} C_i X_i \le \bar{C}, \tag{5.11}$$

where
$\bar{C}$ = the cost level from the solution with (5.3)–(5.4).

Even with this restriction, the spatial layout is still quite important. Ignoring the layout of the harvest could cause the 95 percent confidence timber output to be underachieved by as much as 112.09 cunits. The solution in figure 5.3a met the cost bound by keeping the harvest close to the roads, but dispersing it much more than in figure 5.3b (especially by using both roads). The solution in figure 5.3b is of interest because it also happens to be a minimum-cost solution for harvesting 4.908 cells, without the chance constraint. The cost bound, with nonuniform costs, clearly does affect the results in table 5.1, because the harvest can be relocated only so much while meeting the cost bound. With $b = 1200$, for example, the harvest in the solution to (5.3)–(5.4) exactly surrounds the roads plus .77 of cell 1. The harvest in cell 1 is the only one available to the (5.8)–(5.11) solution to relocate and still meet the area (5.9) and cost (5.11) constraints. Thus, the difference between b and $\hat{b}$ is only 29.41 cunits.

The bottom section of table 5.1 gives the numerical results for the second cost modification; figure 5.4 gives results similar to figures 5.2 and 5.3 (again with $b = 400$) for this cost structure. In figure 5.4a, obtained from formulation (5.3)–(5.4), the harvest is dispersed along the river as expected. In figure 5.4b, obtained from formulation (5.8)–(5.11), the harvest still stays near the river to meet the cost bound (b) but is less dispersed. From table 5.1, the spatial layout is shown to be slightly less critical than in the first cost modification; if one ignored the spatial lay-

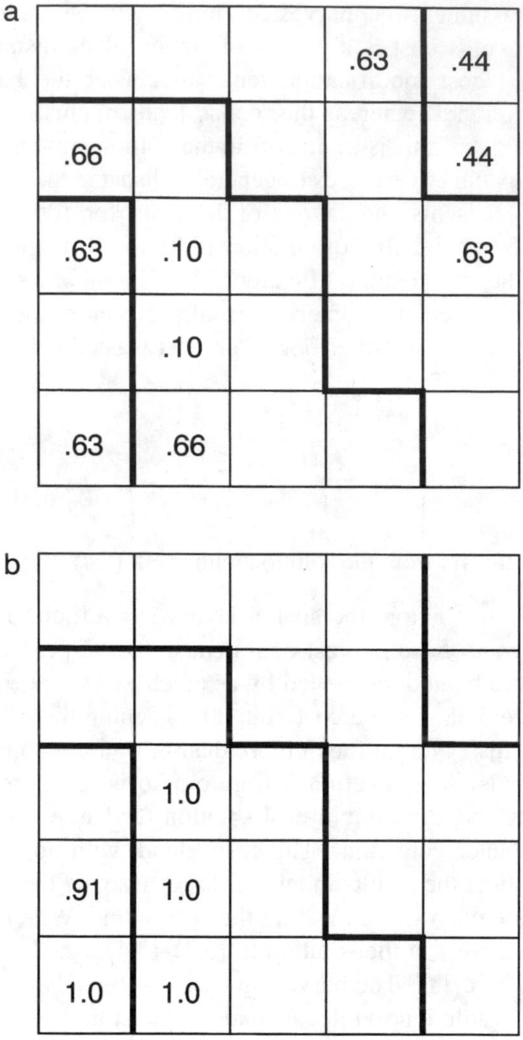

FIGURE 5.3
Optimal proportions of cells harvested for chance-constrained cost minimization (a) and worst-case harvest pattern (b), with costs based on roads for transportation and $b = 400$ (blank indicates zero harvest).

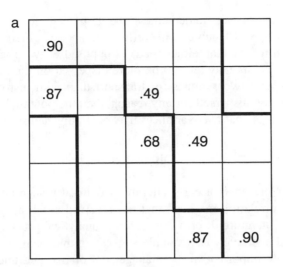

a

.90				
.87		.49		
		.68	.49	
			.87	.90

b

		1.0		
			1.0	
		.20	1.0	
			1.0	1.0

FIGURE 5.4
Optimal proportions of cells harvested for
chance-constrained cost minimization (a) and
worst-case harvest pattern (b), with costs based
on river for transportation and $b = 400$ (blank
indicates zero harvest).

out, they would at most underachieve the 95 percent confidence timber output target by 92.08 cunits. This reflects the fact that there is only one river (though longer than either road) as opposed to two roads. Put another way, harvests may tend to be more dispersed on the basis of cost differentials with two avenues for transport than with just one. If harvests are already dispersed for any reason, then the potential importance of accounting for spatial covariances may be diminished somewhat.

Discussion

This chapter presents an early attempt at optimally accounting for the spatial autocorrelation in yields that has received much attention in recent forestry literature. It provides a simple analytical procedure for this purpose, and a simple case example suggesting that such an accounting could be quite important in the management of risk and uncertainty in forest ecosystems. The implications of spatial autocorrelation between yields on resource tradeoffs may be rather surprising. For example, in comparing the results in figures 5.2, 5.3, and 5.4, note the interaction between the cost structure and the spatial covariances. As the cost structure encourages more clustered harvests, the optimal solution accepts increased total–output variance and harvests more cells to compensate, so as to take advantage of the lower-cost options. If other considerations encourage clustering of harvests, a similar tradeoff might emerge. Spatial autocorrelations, of course, might also be present for many different outputs and ecosystem conditions, not just timber. A wildlife case is discussed in chapter 6.

Chapter 6

A GEOMETRIC WILDLIFE MODEL
WITH SPATIAL AUTOCORRELATION
AND HABITAT CONNECTIVITY

As discussed in chapters 2 and 3, it is commonly predicted that wildlife population establishment and persistence are lower in fragmented habitats than in well-connected habitats, resulting in more susceptibility to extinction (Diamond, 1976; Fahrig and Merriam, 1985; Burkey, 1989; Tilman et al., 1994). In addition to this demographic extinction pressure, there is concern that wildlife populations are vulnerable to environmental stresses (such as fire, extreme weather events, and disease) that have varying magnitudes of spatial covariance (Simberloff and Abele, 1976; den Boer, 1981; Goodman, 1987; Quinn and Hastings, 1987). Persistence time may actually be longer in fragmented landscapes because they spread the risk of environmental stress among subdivided populations (Fahrig and Paloheimo, 1988). The need for connectivity to minimize demographic extinction pressures thus suggests that a desirable spatial layout of habitat fragments would involve highly clumped patches, whereas the possibility of spatially autocorrelated ruinous events suggest that some degree of habitat spreading may be advantageous. This chapter provides a formulation for mathematically capturing both of these considerations and explores some mathematical (nonlinear) programming approaches for finding an optimal balance in this ecological tradeoff (see Hof and Flather, 1996).

This chapter was adapted largely from J. Hof and C. H. Flather, Accounting for connectivity and spatial correlation in the optimal placement of wildlife habitat, *Ecological Modelling* 88 (1996): 143–155, with permission from the publisher, Elsevier Science.

Theory

Connectivity

As in chapters 2 and 3, we assume that wildlife disperse in a random fashion. There is then some probability that a given habitat area near any other habitat area is connected, and this probability diminishes as the distance increases between the two habitat areas. With many habitat areas, the probability of a given area being connected is a function of the number of other habitat areas nearby and the distances to them. We assume, as before, that the probability of each area being connected to a group of areas is the joint probability that the area is connected to *any* (not all) of the areas in the group. We also assume independence between the individual connectivity probabilities. Thus, the joint probability (PR_i) of each patch i being connected is

$$PR_i = 1 - \left[\prod_{j=1}^{M} (1 - pr_{ij}) \right] \quad \forall\, i, \tag{6.1}$$

where pr_{ij} is the probability that patch i is connected to patch j (pr_{ii} set to 0). Presumably, pr_{ij} is smaller the farther patch j is from patch i. Equation (6.1) calculates the joint probability that patch i is not connected to any of the $j = 1, \ldots, M$ ($j \neq i$) patches, and then calculates PR_i as the converse of that joint probability. At some distance, the probability of two habitat areas being connected is effectively 0. Thus, when an area is retained as habitat, it has a certain probability of being connected, which is determined by the number and location of other habitat areas, and it also contributes to the probability of other areas being connected in an equivalent manner. We discuss specific functional relationships between pr_{ij} and interpatch distance in the case example.

As in chapters 2 and 3, we assume that habitat is used only to the degree that it is connected, so the expected population in the ith patch, S_i, is

$$S_i = PR_i a_i A_i \tag{6.2}$$

and the expected value of the total population $E(S)$ is

$$E(S) = \sum_{i=1}^{M} PR_i a_i A_i, \tag{6.3}$$

where
a_i = the expected density of individuals in perfectly connected habitat in the ith patch and
A_i = the size of the ith patch.

Equation (6.3) calculates S_i for each patch as the expected population of a perfectly connected patch $(a_i A_i)$ times the probability that it is connected (PR_i), and then sums across patches to obtain $E(S)$. We initially assume that the a_is are fixed constants, but will also investigate alternative formulations that account for the influence of patch size and shape.

Spatial Autocorrelation

A number of investigators have noted that populations across different patches of habitat are spatially autocorrelated (Gilpin, 1987; Fahrig and Merriam, 1994). Varying degrees of synchrony in population dynamics occur because distance often determines the commonality of random influences (such as weather and fire) on populations, including influences that are directly affected by population connectivity (such as disease and genetic variation). Applying the standard definition of covariance to any two patch populations implies

$$\sigma_{ij}^2 = \rho_{ij}\sigma_i\sigma_j, \tag{6.4}$$

where

$\sigma_{ij}^2 =$ the covariance between the population in patch i and the population in patch j,

$\rho_{ij} =$ the correlation between the population in patch i and the population in patch j, and

$\sigma_i, \sigma_j =$ the standard deviations of the populations in patches i and j, respectively.

Then, the total population variance, $V(S)$, is

$$V(S) = \sum_{i=1}^{M}\sum_{j=1}^{M} \rho_{ij}\sigma_i\sigma_j. \tag{6.5}$$

If the ρ_{ij} are inversely related to the distance between patch i and patch j, then spreading of the patches could be desirable because it reduces the pairwise correlations. This reduces the variance of the total population, thus reducing the probability of a catastrophically low population, all other things (especially $E(S)$) being equal.

There is also typically some relationship between each σ_i and S_i. For convenience, we assume a fixed coefficient of variation for the population of each patch, implying

$$\sigma_i = \Psi S_i \qquad \forall\, i, \tag{6.6}$$

where Ψ is a fixed constant. Thus, by (6.3), (6.5) can be written as

$$V(S) = \sum_{i=1}^{M} \sum_{j=1}^{M} \rho_{ij}(\Psi PR_i a_i A_i)(\Psi PR_j a_j A_j). \qquad (6.7)$$

Specific relationships between the ρ_{ij} and interpatch distance are discussed in the case example.

Chance Maximization

Both connectivity and spatially autocorrelated environmental disturbances can be captured with a mathematical statement of a given confidence level for total population:

$$B = E(S) + \delta V(S)^{1/2}, \qquad (6.8)$$

where

δ = the standard deviate for the given confidence level (a z-value when S is normally distributed, as assumed here) and

B = the population associated with the given confidence level.

Thus, for example, we can calculate the population that we are 80 percent ($\delta = -.84$) confident in (using (6.3) and (6.7)):

$$B = \sum_{i=1}^{M} PR_i a_i A_i - .84 \left[\sum_{i=1}^{M} \sum_{j=1}^{M} \rho_{ij}(\Psi PR_i a_i A_i)(\Psi PR_j a_j A_j) \right]^{1/2}.$$
$$(6.9)$$

It is clear that the size and location of the habitat patches affect B through both the mean and the variance of S in (6.8), as well as in (6.9). We assert that it would be desirable to maximize B with a selected δ, subject to resource limitations (it would also be reasonable to fix B and minimize δ). The remainder of this chapter investigates mathematical programming approaches to this problem, which tie the PR_i and ρ_{ij} variables to specific patch layouts and then optimize those layouts to maximize B.

Optimization

We link the ρ and PR variables to habitat patch layouts using geometric shapes, first circles and then rectangles. Problems with circular patches have a simpler formulation, but the rectangular patches afford more flexibility in terms of patch shape (square versus long and narrow, etc.). Choice variables are established to define location and size (and shape in the case of the rectangles) of habitat patches from which distances can be calculated in the mathematical programs. As in chapter 3, it is necessary to prespecify the maximum number of habitat patches (M).

Circles

As in chapter 3, we locate the circular habitat patches with their centers in a system of rectangular (x,y) coordinates. The size of each circle is then characterized by the radius, and the distance between any two circles is simply the distance between their centers minus the sum of their radii. A mathematical (nonlinear) program to maximize B with circular patches can thus be formulated as follows:

Maximize

$$B = E(S) + \delta V(S)^{1/2}, \tag{6.10}$$

subject to

$$E(S) = \sum_{i=1}^{M} \mathrm{PR}_i a_i A_i, \tag{6.11}$$

$$V(S) = \sum_{i=1}^{M} \rho_{ii}(\Psi \mathrm{PR}_i a_i A_i)^2$$

$$+ \sum_{i=1}^{M} \sum_{j>i} 2\rho_{ij}(\Psi \mathrm{PR}_i a_i A_i)(\Psi \mathrm{PR}_j a_j A_j), \tag{6.12}$$

$$\mathrm{PR}_i = 1 - \left[\prod_{j=1}^{M}(1 - \mathrm{pr}_{ij}) \right] \quad \forall\, i, \tag{6.13}$$

$$\mathrm{pr}_{ij} = f(D_{ij}) \quad \forall\, i, \quad j > i, \tag{6.14}$$

$$\rho_{ij} = g(D_{ij}) \quad \forall\, i, \quad j > i, \tag{6.15}$$

$$\sum_{i=1}^{M} A_i \leq \bar{L}, \tag{6.16}$$

$$x_i \geq r_i \quad \forall\, i, \tag{6.17}$$

$$y_i \geq r_i \quad \forall\, i, \tag{6.18}$$

$$x_i + r_i \leq \bar{X} \quad \forall\, i, \tag{6.19}$$

$$y_i + r_i \leq \bar{Y} \quad \forall\, i, \tag{6.20}$$

$$A_i = \pi r_i^2 \quad \forall\, i, \tag{6.21}$$

$$D_{ij} = \left[(x_i - x_j)^2 + (y_i - y_j)^2\right]^{1/2} - (r_i + r_j) \qquad \forall\, i, \quad j > i,$$

$$(6.22)$$

$$D_{ij} \geq 0 \qquad \forall\, i, \quad j > i, \qquad\qquad (6.23)$$

$$r_i \geq \bar{R} \qquad \forall\, i, \qquad\qquad\qquad\qquad (6.24)$$

where

f and g = distance-dependent functions (possible specific functions are discussed later in the chapter),

D_{ij} = the distance between circle i and circle j,

$\bar{L}$ = the amount of habitat area that can be retained,

$\bar{X}$ = the east–west dimension of the problem space,

$\bar{Y}$ = the north–south dimension of the problem space,

x_i = the x-coordinate of the center of the ith circular habitat patch

y_i = the y-coordinate of the center of the ith circular habitat patch,

r_i = the radius of the ith circular habitat patch,

$\bar{R}$ = the minimum radius for each habitat circle,

and all other variables are as previously defined.

The optimization process will choose levels of x_i, y_i, and r_i that size and locate the habitat circles so as to maximize B. Equation (6.13) merely repeats equation (6.1) and calculates the joint probability of connectivity for each patch i. Equation (6.14) calculates the pairwise probability of connectivity for each i and j pair as a function of distance. Similarly, equation (6.15) calculates the correlation for each i and j pair as a function of distance. Equation (6.16) limits the total amount of retained habitat to $\bar{L}$. Equations (6.17)–(6.20) keep the circles of habitat within the problem space, which is assumed to be rectangular with dimensions $\bar{Y}$ by $\bar{X}$. Equation (6.21) calculates the area of each circle, and equation (6.22) calculates the distance between each pair of circles. Equation (6.23) prevents the circles from overlapping. Equation (6.24) sets the minimum radius for each habitat circle. If it is desired to allow radii (and thus circle areas) to go to 0, the contribution to population variance is automatically removed in equation (6.9) (see also equation (6.5)). D_{ij} would have to be multiplied by $A_i A_j / (A_i A_j + \varepsilon)$ in equation (6.14), where ε is an arbitrarily small constant, in order to remove any contribution of a zero-area circle to connectivity. This would allow selection of the number of habitat patches, within the maximum allowed number, M. If the number of patches is to be prespecified (as M), then $\bar{R}$ should be set at the minimum size that functions as a patch in the model in terms of carrying capacity, connectivity, and covariance.

In the formulations up to this point, we have treated the a_is as fixed constants. For many species, the density of population is not a simple linear function of habitat area. Often, edge habitats—habitats near the boundary of the patch—have either unsuitable microclimates or harbor predators and competitors that can reduce the population. Consequently, as the proportion of edge habitat increases relative to patch area, the expected value of the population should decline. To account for this phenomenon, we defined a buffer distance b from the habitat patch edge, inside of which edge-associated population reduction factors cease to affect population density. We could then penalize a_i as follows (to define α_i):

$$\alpha_i = a_i(A_T - A_b)^\gamma \qquad 0 \le \gamma \le 1, \qquad (6.25)$$

where A_T is total patch area, A_b is the area of the buffer, and γ reflects the degree to which species can survive and reproduce in the edge habitats. If the edge habitats are totally unsuitable, then $\gamma = 1$ and the effective habitat area (AA_i) could be calculated as

$$AA_i = \pi(r_i - b)^2 \qquad \forall\, i \qquad (6.26)$$

and AA_i would replace A_i in equation (6.9), but not (6.16), which constrains the total habitat area, and (6.21), which defines the A_i. The a_i would still be fixed because the nonlinearity is accounted for in calculating AA_i. Because the shape of the circles is invariant, this penalizes only small patches in terms of habitability. We next turn to a formulation that uses rectangles, so that the shape of the patches is more variable.

Rectangles

Our approach with rectangles of habitat is similar, but equations (6.17)–(6.24) must be replaced to account for the different geometry. For convenience, we locate the ith rectangle by the coordinates (x_i°, y_i°) of its southwest (lower-left-hand) corner. The size and shape of each habitat rectangle is then determined by two choice variables (x_i^* and y_i^*) specifying its x and y dimensions.

Calculating the distance between rectangles is more complicated than with circles because of the variable shape. Let us define the x and y vectors between the closest points of two rectangles (i and j) as Δx and Δy. Then note that the distance between i and j is always

$$D_{ij} = \sqrt{\Delta x^2 + \Delta y^2}, \qquad (6.27)$$

which is Δx if $\Delta y = 0$, Δy if $\Delta x = 0$, and the hypotenuse of the triangle formed by Δx and Δy if $\Delta x > 0$ and $\Delta y > 0$.

Now, looking first at the x vector, there are three cases to consider:

(a) rectangle i is completely to the left of rectangle j, so $\Delta x = x_j^\circ - (x_i^\circ + x_i^*)$;
(b) rectangle i is completely to the right of rectangle j, so $\Delta x = x_i^\circ - (x_j^\circ + x_j^*)$; and
(c) rectangles i and j are at least partially above/below each other, so $\Delta x = 0$.

Note that in case (a) $x_i^\circ - (x_j^\circ + x_j^*)$ is negative and that in case (b) $x_j^\circ - (x_i^\circ + x_i^*)$ is negative.

Similarly, looking at the y vector, there are three cases to consider:

(d) rectangle i is completely below rectangle j, so $\Delta y = y_j^\circ - (y_i^\circ + y_i^*)$;
(e) rectangle i is completely above rectangle j, so $\Delta y = y_i^\circ - (y_j^\circ + y_j^*)$; and
(f) rectangles i and j are at least partially left/right of each other, so $\Delta y = 0$.

Note that in case (d) $y_i^\circ - (y_j^\circ + y_j^*)$ is negative and that in case (e) $y_j^\circ - (y_i^\circ + y_i^*)$ is negative.

It is useful to define the following instrumental variables:

$$DX_{ij}^1 = .5\sqrt{(x_j^\circ - x_i^\circ - x_i^*)^2} + .5(x_j^\circ - x_i^\circ - x_i^*)$$

$$DX_{ij}^2 = .5\sqrt{(x_i^\circ - x_j^\circ - x_j^*)^2} + .5(x_i^\circ - x_j^\circ - x_j^*)$$

$$DY_{ij}^1 = .5\sqrt{(y_j^\circ - y_i^\circ - y_i^*)^2} + .5(y_j^\circ - y_i^\circ - y_i^*)$$

$$DY_{ij}^2 = .5\sqrt{(y_i^\circ - y_j^\circ - y_j^*)^2} + .5(y_i^\circ - y_j^\circ - y_j^*)$$

For example, $DX_{ij}^1 = 0$ if $x_j^\circ - x_i^\circ - x_i^*$ is negative, but is $x_j^\circ - x_i^\circ - x_i^*$ otherwise. Case (a) implies that $DX_{ij}^1 > 0$ and $DX_{ij}^2 = 0$; case (b) implies that $DX_{ij}^1 = 0$ and $DX_{ij}^2 > 0$; and case (c) implies that $DX_{ij}^1 = 0$ and $DX_{ij}^2 = 0$. With these instrumental variables, it is then possible to calculate the distance between rectangle i and rectangle j as

$$D_{ij} = \left[(DX_{ij}^1)^2 + (DX_{ij}^2)^2 + (DY_{ij}^1)^2 + (DY_{ij}^2)^2\right]^{1/2}, \quad (6.28)$$

with the restriction that

$$DX_{ij}^1 + DX_{ij}^2 + DY_{ij}^1 + DY_{ij}^2 \geq \mu, \quad (6.29)$$

where μ is an arbitrarily small positive constant, to prevent the two rectangles from overlapping. The constant μ must be selected to exceed the precision of the instrumental variable calculations, as it works by forcing at least one instrumental variable to be greater than 0. Equation (6.28) is simply a restatement of equation (6.27), using the instrumental variables to zero out the incorrect calculations of Δx and Δy so that (6.28) is general to all combinations of cases (a), (b), and (c) with cases (d), (e), and (f).

A mathematical (nonlinear) program to maximize B with rectangular patches can thus be formulated by replacing (6.17)–(6.24) in the circle formulation with the following:

$$x_i^\circ \geq 0 \qquad \forall i, \tag{6.30}$$

$$y_i^\circ \geq 0 \qquad \forall i, \tag{6.31}$$

$$x_i^\circ + x_i^* \leq \bar{X} \qquad \forall i, \tag{6.32}$$

$$y_i^\circ + y_i^* \leq \bar{Y} \qquad \forall i, \tag{6.33}$$

$$A_i = x_i^* \cdot y_i^* \qquad \forall i, \tag{6.34}$$

$$D_{ij} = \left[(DX_{ij}^1)^2 + (DX_{ij}^2)^2 + (DY_{ij}^1)^2 + (DY_{ij}^2)^2 \right]^{1/2} \quad \forall i, \quad j > i, \tag{6.35}$$

$$DX_{ij}^1 + DX_{ij}^2 + DY_{ij}^1 + DY_{ij}^2 \geq \mu \qquad \forall i, \quad j > i, \tag{6.36}$$

$$x_i^* \geq \bar{Q} \qquad \forall i, \tag{6.37}$$

$$y_i^* \geq \bar{Q} \qquad \forall i, \tag{6.38}$$

Equations (6.30)–(6.33) keep the rectangles of habitat within the problem space, as previously defined. Equation (6.34) calculates the area of each habitat rectangle. Equation (6.35) calculates the distances between rectangles, and equation (6.36) prevents overlaps, as just described. $\bar{Q}$ is the minimum size of each dimension of each rectangle of habitat. The same adjustment to equation (6.14) (as in the circle formulation) would be necessary if it is desired to set $\bar{Q} = 0$. All other variables are as previously defined.

In order to account for unusable buffer areas near edges, as discussed for the circle formulation (and again assuming that the buffer areas are completely unsuitable as habitat), we could define

$$AA_i = (x_i^* - b)(y_i^* - b) \qquad \forall i \tag{6.39}$$

and replace A_i with AA_i in equation (6.9). This penalizes long, narrow shapes as well as small sizes of rectangular habitat patches. It should be noted that if only shape is to be penalized, this would be possible by replacing a_i with α_i, defined in a manner such as

$$\alpha_i = a_i \left(\frac{4\sqrt{A_i}}{2x_i^* + 2y_i^*} \right)^\lambda \tag{6.40}$$

(see Austin, 1984), which would penalize the a_i for shapes as they deviate from squares, at a rate determined by λ ($0 \leq \lambda \leq 1$).

An Example

The Problem

In order to construct a case example, we scaled the spatial optimization problem to the ecology of a hypothetical species that defends a 1-ha territory and each territory represents a single breeding pair. These life history attributes are not entirely arbitrary but are characteristic of an avian, habitat-interior specialist (see Temple and Cary, 1988). We defined patch connectivity as the probability of individuals successfully immigrating from patch i to patch j. Patch connectivity is thus a function of distance between patches (measured as the minimum edge-to-edge distance), the dispersal capability of the species, and the harshness of the interpatch environment. A continuous function for pr_{ij} that declines monotonically with distance and approaches 0 asymptotically is given by

$$\mathrm{pr}_{ij} = \mathrm{pr}^0 - \mathrm{pr}^0(1 - \theta^{D_{ij}})^\beta, \tag{6.41}$$

where β reflects species dispersal capability and sets a threshold distance beyond which the probability of successfully colonizing a patch declines rapidly, and θ reflects the harshness of the interpatch environment, which affects the rate of decline. The parameter pr^0 indicates the probability of connectivity when $D_{ij} = 0$.

The formulations also require a function that relates patch autocorrelation (ρ_{ij}) to distance (D_{ij}). Clearly, the autocorrelation between patch populations increases as the distance between patches declines. However, the autocorrelation among patches is also affected by the type of environmental disturbance agents that affect long-term population persistence. For example, population dynamics in fragmented habitats would have an inherently different ρ_{ij} structure if the disturbance agent is a disease transmitted by contact among individuals rather than a disturbance agent unaffected by patch population interactions (such as severe drought). We represent ρ_{ij} as a function of distance between patches as

$$\rho_{ij} = \rho^0 - \rho^0(1 - \omega^{D_{ij}})^\tau, \tag{6.42}$$

where τ reflects a threshold distance beyond which the autocorrelation between patches declines rapidly and ω reflects the rate at which the spatial covariance among patches decreases with distance. Both τ and ω are disturbance-specific. The parameter ρ^0 indicates the correlation when $D_{ij} = 0$.

A square problem space of 10,000 ha was then defined and the unit distance within the problem space was set to 100 m, which is the linear measure of one side of a square territory 1 ha in size (so the problem space is 100×100 units and one area unit (1 ha) is one animal territory). We set a confidence level for equation (6.9) at 80 percent, and assumed that patch populations have a 0.5 coefficient of variation. We assumed that only 2000 ha of habitat could be retained among four habitat patches. For simplicity, it was assumed that all patches are large enough to support at least one breeding pair, and all function as patches in terms of connectivity and covariance. Thus, in terms of the formulations previously presented, the following parameter settings were used:

$M = 4$

$a_i = 2$

$\delta = -.84$

$\Psi = .5$

$\bar{L} = 2000,$

$\bar{X} = 100,$

$\bar{Y} = 100,$

$\bar{R} = .56419,$ and

$\bar{Q} = 1.$

We set $\rho^0 = \mathrm{pr}^0 = .9$ so that adjacent patches of habitat are highly connected and highly correlated, but still distinguished as separate patches. For demonstration purposes, both β and τ were set at 50, and θ and ω were varied between values of 0.75, 0.85, and 0.95.

Results

For all the solutions presented, solver tolerances were set such that feasibility is ensured within 1×10^{-6}. Multiple starts were used in an effort to ensure global optimality, and all solutions presented appear to be global optima, but only local optimality can be absolutely ensured.

We initially solved the circular patch model with no size penalty ($\gamma = 0, b = 0$). Figure 6.1a presents the solution with parameters set to reflect a species whose dispersal is not strongly inhibited by the interpatch environment (that is, $\theta = 0.95$, so pr_{ij} decays slowly with interpatch distance) and where the disturbance agents tend to be rather local

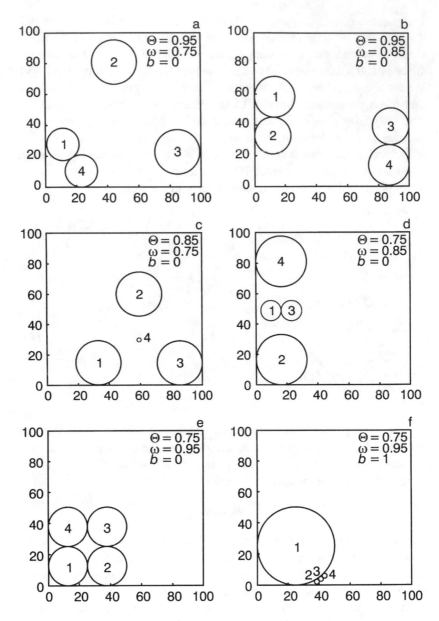

FIGURE 6.1

Solutions of the circular patch formulation with varying levels of θ and ω. Layouts a–e have no size penalty, whereas f implements an edge buffer of unsuitable habitat that penalizes patch size.

TABLE 6.1
Connectivity Probabilities, Correlations, and Objective Function Components
for Figures 6.1a–f.

	Fig. 6.1a	Fig. 6.1b	Fig. 6.1c	Fig. 6.1d	Fig. 6.1e	Fig. 6.1f
pr_{12}	.8996	.9000	.6170	.8596	.9000	.9000
pr_{13}	.8876	.8752	.6172	.9000	.8288	.9000
pr_{14}	.9000	.8222	.8946	.8596	.9000	.9000
pr_{23}	.8983	.8729	.6171	.8596	.9000	.9000
pr_{24}	.8875	.8752	.8946	.0052	.8287	.9000
pr_{34}	.8996	.9000	.8946	.8596	.9000	.9000
PR_1	.9989	.9978	.9845	.9980	.9983	.9990
PR_2	.9988	.9984	.9845	.9804	.9983	.9990
PR_3	.9988	.9984	.9845	.9980	.9983	.9990
PR_4	.9989	.9978	.9988	.9804	.9983	.9990
ρ_{12}	.00083	.9000	.0545	.9000	.9000	.9000
ρ_{13}	.00004	.0095	.0544	.9000	.9000	.9000
ρ_{14}	.90000	.0029	.5016	.9000	.9000	.9000
ρ_{23}	.00027	.0088	.0545	.9000	.9000	.9000
ρ_{24}	.00004	.0095	.5012	.2334	.9000	.9000
ρ_{34}	.00082	.9000	.5012	.9000	.9000	.9000

Objective Function

B	3034.6	2831.8	2931.7	2557.7	2380.1	2115.7
$E(S)$	3995.4	3992.3	3938.3	3931.4	3993.1	3647.3
$V(S)^{1/2}$	1143.8	1381.5	1198.4	1635.4	1920.2	1823.3

(that is, $\omega = 0.75$, resulting in spatial autocorrelations that decrease rapidly with distance). Two patches of habitat are clustered together, with the other two spread out nearly as far as possible. Note in table 6.1 that the pairwise and joint probabilities of connectivity are still quite high; throughout the analysis, the solutions do not disperse habitat to the point that connectivity is severely sacrificed. At the same time, when an animal is capable of migrating across nontrivial distances, there is clearly an advantage in dispersing the habitat to some degree to reduce total population variance. And, given the particular formulations developed, there are often ways to disperse the habitat and reduce total population variance without serious adverse effects on overall connectivity. The autocorrelations between all patches except 1 and 4 are quite low (table 6.1).

If ω is increased to 0.85, implying that spatial autocorrelations do not decrease quite as rapidly with distance, then patches cluster into two pairs (figure 6.1b), which increases the connectivity relative to figure 6.1a, but still disperses the pairs widely. In table 6.1, comparing the first with the second solution, $\omega = 0.85$ causes the solution population to decrease and the population standard deviation to increase.

In figure 6.1c, ω is returned to 0.75 and θ is decreased to 0.85, indicating an interpatch matrix that is more resistant to successful dispersal. This results in the patches being arranged in an equilateral triangle, with a "stepping stone" patch in the middle. In this type of solution, the assumption that any patch with $r_i \geq \bar{R}$ is fully functional in terms of connectivity (and autocorrelation) is critical. In table 6.1, this solution has some pairwise connectivity probabilities that are much lower than in previous solutions, but the joint connectivity probabilities are still quite high because of the location of patch 4. For this connectivity, the solution endures much higher autocorrelations between patch 4 and the other patches. It is still possible to keep the covariance small, however, because $\omega = 0.75$ (table 6.1).

In figure 6.1d, θ is decreased to 0.75 and ω increased (again) to 0.85. These parameters reflect a very harsh interpatch environment and disturbance factors that affect a large enough area to cause spatial autocorrelations to decline only moderately with distance. The solution spreads patches 2 and 4 rather widely, but places patches 1 and 3 in between for connectivity. The solution endures high autocorrelations between all patch pairs except 2 and 4, resulting in a high population standard deviation (table 6.1). Looking across solutions in table 6.1, it is clear that population levels with 80 percent confidence generally decrease with lower migration capabilities (caused by a harsh interpatch matrix) and large disturbance factors.

In figure 6.1e, θ is kept at 0.75 and ω is increased to 0.95. These parameters again reflect an animal that is less capable of crossing nonhabitat area, but now with disturbance factors that are large enough to cause spatial autocorrelations to decrease slowly with increasing distance. This implies that retaining connectivity will require close distances, and that dispersing habitat patches gains little in reducing population variances. The resulting solution is not surprising: four equal-sized patches are grouped as close together as possible. We did confirm that it is still optimal to retain these four patches as opposed to one large patch, because some variance reduction is still obtained (because $\rho^0 = .9$) and the joint connectivity probabilities are all very near 100 percent (table 6.1).

This is not the case if a buffer 100 m wide ($\gamma = 1$, $b = 1$) is removed from the calculated usable habitat in each patch in equation (6.9) (figure 6.1f). Note that $\bar{R}$ was increased to 1.56419 to retain a minimum usable habitat for one species pair in each patch. Here, θ and ω are left at 0.75 and 0.95, respectively, but eliminating the buffer from usable habitat penalizes small patches, which results in one large patch and three minimum-sized patches placed adjacent to the large one. The buffer is about 15 percent of the habitat area as configured in figure 6.1e, but is about 10 percent in figure 6.1f, which is enough improvement to alter the optimal layout as observed. This demonstrates the difference that one might expect between edge-neutral species (figure 6.1e) and habitat-interior specialists (figure 6.1f) that cannot survive close to the edge (see Margules et al., 1994). Because circles are invariant with regard to shape, this buffer effect penalizes only size. Figures 6.2a–f and table 6.2 demonstrate the rectangular-patch formulation, which allows some variation in shape.

Figure 6.2a presents the solution to the rectangular-patch formulation with $\theta = 0.85$ and $\omega = 0.75$ (as in figure 6.1c) and without the edge

TABLE 6.2
Connectivity Probabilities, Correlations, and Objective Function Components
for Figures 6.2a–f.

	Fig. 6.2a	Fig. 6.2b	Fig. 6.2c	Fig. 6.2d	Fig. 6.2e	Fig. 6.2f
pr_{12}	.6117	.6045	.8985	.8821	.9000	.9000
pr_{13}	.8980	.8995	.8985	.8821	.9000	.9000
pr_{14}	.6117	.6045	.000015	.0007	.9000	.8997
pr_{23}	.8982	.8996	.00002	.0066	.9000	.9000
pr_{24}	.6046	.6028	.8985	.8821	.9000	.9000
pr_{34}	.8982	.8996	.8985	.8821	.9000	.9000
PR_1	.9846	.9843	.9897	.9861	.9990	.9990
PR_2	.9844	.9842	.9897	.9862	.9990	.9990
PR_3	.9989	.9990	.9897	.9862	.9990	.9990
PR_4	.9844	.9842	.9897	.9861	.9990	.9990
ρ_{12}	.0530	.0510	.9000	.9000	.9000	.9000
ρ_{13}	.5995	.7097	.9000	.9000	.9000	.9000
ρ_{14}	.0530	.0510	.0098	.0825	.9000	.9000
ρ_{23}	.6104	.7176	.0116	.2613	.9000	.9000
ρ_{24}	.0510	.0506	.9000	.9000	.9000	.9000
ρ_{34}	.6104	.7176	.9000	.9000	.9000	.9000
Objective Function						
B	2933.9	2486.7	2744.7	2270.7	2381.8	2088.9
$E(S)$	3937.9	3336.3	3958.8	3307.2	3996.0	3601.0
$V(S)^{1/2}$	1195.1	1011.5	1445.3	1233.9	1921.6	1800.2

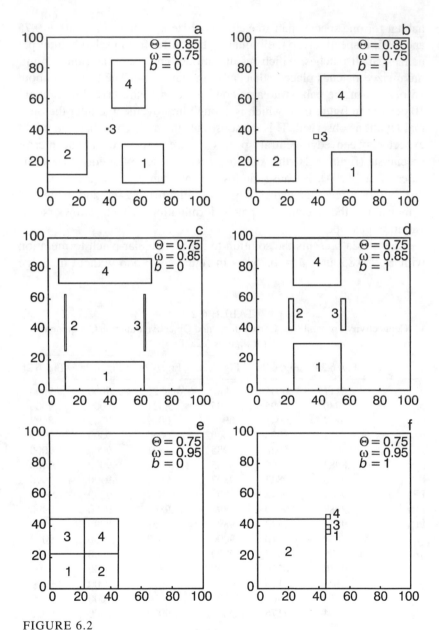

FIGURE 6.2
Solutions of the rectangular patch formulation with varying levels of θ and ω. Layouts a, c, and e have no size or shape penalty, whereas b, d, and f implement an edge buffer of unsuitable habitat that penalizes patch size and shape.

buffer penalty. The results are very similar to those in figure 6.1c, but with patch shapes that are somewhat more elongated. The equivalent solution with the buffer ($\gamma = 1$, $b = 1$) removed from usable habitat in equation (6.9) results in equalizing the three larger patch areas, increasing the interior patch size to meet the new constraint, and reshaping the larger patches into squares (figure 6.2b). Note that $\bar{Q}$ was increased to 3.0 to retain a minimum usable habitat for one species pair in each patch. In table 6.2, the effect of removing the edge buffer is to scale back the objective function (B) and its components $E(S)$ and $V(S)^{1/2}$, but to leave the connectivity probabilities and autocorrelations almost unchanged.

The solution to the rectangular-patch formulation with $\theta = 0.75$ and $\omega = 0.85$ (as in figure 6.1d) and no buffer penalty makes considerable use of the shape flexibility in the rectangular-patch model, increasing the objective function by about 7 percent (figure 6.2c, table 6.2). Two long, narrow patches are used to connect two larger patches while at the same time keeping the distance between the larger patches quite large (and correlation quite small, see table 6.2). The assumption on $\bar{Q}$ is again critical because the model implicitly assumes that patches 2 and 3 are wide enough to serve as corridors connecting patches 1 and 4. Incidentally, patches 2 and 3 are also connected by patches 1 and 4, and patches 1 and 4 are apparently elongated to reduce the correlation between 2 and 3. An equivalent solution with the adjustments to remove the buffer from usable habitat squares up patches 1 and 4 and widens patches 2 and 3 to meet the new $\bar{Q}$ level (figure 6.2d). Because of the problem space limits and the geometry of patches 1 and 4, this also implies shortening patches 2 and 3, and shortening the distances between most of the patches. The solution values are quite different for the parameter settings depicted in figures 6.2c and 6.2d (table 6.2).

The solution to the rectangular-patch formulation with $\theta = 0.75$, $\omega = 0.95$, and no buffer penalty is similar to the circular-patch formulation. In fact, the advantages of compaction cause the optimal solution to create nearly square shapes even without the buffer penalty (compare figures 6.2e and 6.1e). When the buffer is removed from the usable habitat (figure 6.2f), however, a solution much like figure 6.1f is obtained.

Discussion

The case example is not intended as a realistic application of the model formulations. Our objective was to mathematically capture the mutually inconsistent ecological concerns of habitat fragmentation and spatial autocorrelation into a single model and show that the optimal balance between these concerns can, in principle, be determined through optimization procedures. The case example solutions do suggest an ecological principle that the best spatial arrangement for habitat often includes a mixed strategy that manages to connect the habitat but still spread it out at the same time (as in figures 6.1d, 6.2c, and 6.2d). Also, it is clear that simple ecological principles are difficult to develop because species with different life history attributes (represented here by dispersal capability) and environmental disturbance agents with different spatial magnitudes both imply radically different optimal layouts.

Chapter 7

PRAGMATIC APPROACHES TO HANDLING RISK AND UNCERTAINTY

The approaches in chapters 5 and 6 require nonlinear formulations and solution procedures. These are not practical in many analysis situations where random technical coefficients are a serious problem (including the models discussed in chapters 2 through 4) and where model size or complexity is substantial. The purpose of this chapter is to provide some pragmatic approaches to accounting for A-matrix randomness that are applicable to large linear programming models with standard solution software (see Hof et al., 1995).

The Problem

Let us assume we wish to solve a general linear program:
    Minimize

$$\sum_{j=1}^{n} c_j X_j, \tag{7.1}$$

subject to

$$\sum_{j=1}^{n} a_{ij} X_j \geq b_i \qquad \forall\, i \tag{7.2}$$

$$X_j \geq 0 \qquad \forall\, j,$$

Most of this chapter is adapted from J. Hof, M. Bevers, and J. Pickens, Pragmatic approaches to optimization with random yield coefficients, *Forest Science* 41(3) (1995): 501–512, with permission from the publisher, the Society of American Foresters.

but the problem is that the a_{ij} are random, with expected values α_{ij} and variances/covariances σ^2_{ijh}. It is well established that simply substituting α_{ij} for a_{ij} and ignoring the randomness can generate misleading results (Hof et al., 1988). The classic approach, discussed in the introduction to part II and in chapters 5 and 6, is the chance-constrained (and related chance maximization) method of Van de Panne and Popp (1963) and Miller and Wagner (1965), which defines a probability, θ_i, that a constraint is met as

$$\text{prob}\left(\sum_{j=1}^{n} a_{ij} X_j \geq b_i \right) \geq \theta_i \qquad \forall\, i, \tag{7.3}$$

but this approach, again, requires nonlinear optimization methods. This limits the size of the problem that can be solved and raises the possibility of local optimality in some cases (see Hof et al., 1992; Weintraub and Vera, 1991).

Postoptimization Calculations

Perhaps the most straightforward approach to accounting for a random A-matrix in a large problem is to replace the a_{ij} with their means (α_{ij}) and solve the problem with standard linear programming software, but then to perform some postoptimization calculations before interpreting the results. Starting with (7.2), the row total

$$\sum_{j=1}^{n} a_{ij} X_j \tag{7.4}$$

for each row i is a linear function of the a_{ij} because the X_j are constant in any given solution. The ith row total thus has an expected value of

$$\sum_{j=1}^{n} \alpha_{ij} X_j \tag{7.5}$$

and a variance of (see Miller and Wagner, 1965)

$$\sum_{j=1}^{n} \sum_{h=1}^{n} X_j X_h \sigma^2_{ijh}. \tag{7.6}$$

For a solution based on the α_{ij} and a given $\tilde{b}$ right-hand side vector, we can specify a single-tailed confidence level for each row, and then calculate the level of resultant row-total attainment for which that confidence applies. Suppose we specify a confidence level of θ_i and the associated tabular value of the standard deviate is δ_i. If the a_{ij} are normally distributed, then the row total (a linear function of a_{ij}) will also be normal and δ_i is the z value or the standard normal deviate. Then, for any optimal solution $\tilde{X}^*$, we can calculate that we are θ-confident that at least

$$\sum_{j=1}^{n} \alpha_{ij} X_j^* + \delta_i \left(\sum_{j=1}^{n} \sum_{h=1}^{n} X_j^* X_h^* \sigma_{ijh}^2 \right)^{1/2} = \beta_i^* \qquad (7.7)$$

will be obtained for the ith row. If the calculated $\tilde{\beta}^*$ vector is not acceptable, another run with the α_{ij} and an altered $\tilde{b}$ vector could be attempted. The power of this approach is that the θ-confidence is correct for the $\tilde{X}^*$ solution and the calculated $\tilde{\beta}^*$ vector. This at least ensures that decision makers (and analysts) get a realistic depiction of the outputs to be expected from the $\tilde{X}^*$ solution. It should be emphasized, however, that the $\tilde{X}^*$ solution is optimal (for example, minimum cost) with regard to the $\tilde{b}$ vector, *not* the $\tilde{\beta}^*$ vector. A solution may exist that has smaller row-total variances than $\tilde{X}^*$ and thus obtains the $\tilde{\beta}^*$ vector with θ-confidence at lower expected cost.

Before proceeding, we should briefly discuss the covariances between technical coefficients within a given constraint.[1] When each constraint in (7.2) is defined for a given time period, all the coefficients in each constraint apply to the same time period, so intertemporal covariances would not be relevant. If different management prescriptions apply to the same analysis area and involve similar management actions, the response coefficients would be expected to be very highly correlated. On the other hand, if two different management prescriptions apply to highly differentiated analysis areas (either by qualitative differences or by distance), then the covariance between the response coefficients might be quite small. The middle ground would involve spatial covariances that are smaller than the coefficient variances, but still large enough to be important. These different circumstances are explored in a case example.

An Example

Suppose we have a 4000-acre tree farm, growing slash pine (*Pinus elliottii*) pulpwood, and we wish to schedule harvests over six 4-year time

---

1. We are considering constraints individually and assume zero between-row covariances.

periods. The initial age structure includes four analysis areas of 1000 acres each, aged 8, 12, 16, and 20 years. The discount rate is set at 4 percent, and assume that we wish to minimize discounted costs ($16 per cord and $200 per acre harvesting and regeneration costs before discounting) in meeting output targets set for each time period. Yield means based on Bennett (1970) are as follows:

Age (yr)	Yield (cords/acre)
16	27.5
20	42.6
24	55.1
28	65.1
30 and older	69.1

Assume that harvesting is not allowed before age 16 and yield means are constant after age 30 (at least within the age limits of our model). All discounting and yield projections are based on the beginning of each time period, and alternative management prescriptions are included for all possible combinations of single and multiple harvests. For simplicity, let us assume that all yields are normally distributed and have a coefficient of variation of 0.5. Furthermore, all yields that result from any harvest on the same analysis area in the same time period are perfectly correlated ($\rho = 1$) even if (under different prescriptions) they involve different-aged timber. This implies a covariance equal to the product of the standard deviations for any two yield coefficients (which implies that the covariance equals the variance for equal yields). The covariances between yields on different analysis areas are assumed to be 0 for reasons made clear later in the chapter. The case example was kept small enough to allow nonlinear optimization with chance constraints, to serve as a standard of comparison.

A typical linear programming formulation for this problem would be as follows:

Minimize

$$C = \sum_{j=1}^{4} \sum_{k=1}^{n_j} c_{jk} X_{jk}, \tag{7.8}$$

subject to

$$\sum_{j=1}^{4} \sum_{k=1}^{n_j} a_{tjk} X_{jk} \geq b_t \qquad \forall \, t, \tag{7.9}$$

$$\sum_{k=1}^{n_j} X_{jk} \leq 1000 \qquad \forall \, j, \tag{7.10}$$

where

C = total discounted cost,

X_{jk} = the number of acres allocated to the kth management prescription for the jth analysis area,

c_{jk} = the discounted cost per acre allocated to X_{jk},

a_{tjk} = the yield in time period t for every acre allocated to the kth management prescription for the jth analysis area, and

b_t = the right-hand side output target for time period t.

Note that the a_{tjk} are random, which causes the objective function coefficients in (7.8) to also be random, because those coefficients typically involve per acre *and per volume* costs for whatever yields result from management prescription X_{jk}. If we simply insert the expected values of the c_{jk} into the objective function and minimize it, the results are equivalent to minimizing the expected value (50 percent confidence level) of the objective function (see Hof et al., 1988). For this example, we take this approach, but it should be noted that a different confidence level for the objective function would be more difficult to accommodate. Therefore, the objective function will hereafter be called $E(C)$, and the c_{jk} are defined as *expected* discounted costs.

We treat the randomness in constraint (7.9) (from the a_{tjk} being random) with a different confidence level. Suppose the decision maker would like to be 80 percent confident of obtaining total yields of 25,000 cords in each time period. The z value associated with (single-tailed) 80 percent confidence is -0.84. The nonlinear programming approach from Van de Panne and Popp (1963) and Miller and Wagner (1965) would thus minimize (7.8) subject to (7.10) and

$$\sum_{j=1}^{4}\sum_{k=1}^{n_j} \alpha_{tjk}X_{jk} - 0.84\left(\sum_{j=1}^{4}\sum_{k=1}^{n_j}\sum_{l=1}^{4}\sum_{m=1}^{n_l} X_{jk}X_{lm}\sigma_{tjklm}^2\right)^{1/2}$$
$$\geq 25,000, \quad (7.11)$$

where

$$n_l = n_j \text{ for } l = j. \quad (7.12)$$

Van de Panne and Popp showed that with all $\delta_i \leq 0$ $(\theta_i \geq \frac{1}{2})$, this is a convex program, so local optimality is not a problem. The optimal solution, obtained with the GRG2 nonlinear algorithm, indicates a cost of $2,871,256 to precisely meet output targets of 25,000 cords with 80 percent confidence in each time period. This solution serves as a standard of comparison for the approaches that follow.

TABLE 7.1
Solutions with Postoptimization Calculations (all units are in 1000s)

	Right-Hand Side in Linear Program	Row Total Mean	Row Total Standard Deviation	Row Total with 80% Confidence
Solution 1 (cost = 2064.56)				
$t = 1$	25	25	12.50	14.50
$t = 2$	25	25	11.44	15.39
$t = 3$	25	25	12.50	14.50
$t = 4$	25	25	12.50	14.50
$t = 5$	25	25	9.50	17.02
$t = 6$	25	25	12.50	14.50
Solution 2 (cost = 3382.26)				
$t = 1$	39	39	19.50	22.62
$t = 2$	39	39	17.33	24.44
$t = 3$	39	39	15.13	26.29
$t = 4$	39	39	13.81	27.40
$t = 5$	39	39	15.20	26.23
$t = 6$	39	39	14.35	26.94
Solution 3 (cost = 3450.23)				
$t = 1$	41	41	20.50	23.78
$t = 2$	39	39	18.49	23.46
$t = 3$	39	39	16.25	25.35
$t = 4$	39	39	14.30	26.99
$t = 5$	39	39	14.00	27.24
$t = 6$	39	39	13.89	27.33

An analysis using the postoptimization calculations previously outlined might proceed as follows. Solution 1 in table 7.1 gives the results of simply running the linear program with the mean coefficients and all six right-hand sides set at 25,000 cords. The row totals that are actually associated with 80 percent confidence, calculated with (7.7), vary across time periods between 14,500 and 17,020, as opposed to the desired 25,000. The model was then successively solved, increasing the right-hand sides until they were all 39,000, at which point row totals with 80 percent confidence were fairly close to 25,000 in time periods 2, 3, 4, 5, and 6 (Solution 2, table 7.1). Another simple adjustment would be to increase the right-hand side in time period 1 to, say, 41,000, which results in Solution 3. At this point, the 80 percent confidence row totals are all fairly close to the desired 25,000, but it should be noted that increasing the right-hand side for time period 1 affects the attainment in other time periods. In most applications, the confidence levels and output targets are rather imprecise, so an approach such as this would often provide acceptable results—certainly better results than the initial run based simply on mean coefficients and the right-hand sides of 25,000.

In viewing table 7.1, it is tempting to suggest a more systematic approach. Solution 1 in table 7.2 provides a starting point. Solution 1 resulted from setting the right-hand sides to 35,500. This number was obtained as a guess by taking 25,000 plus 0.84 times the coefficient of variation times 25,000. In cases without a fixed coefficient of variation, an average standard deviation across all A-matrix coefficients might be used instead. In any case, this is just a rough guess because we are trying to anticipate the standard deviation of the row total, based on the standard deviations of the A-matrix coefficients. The next step is Solution 2 in table 7.2, where the right-hand sides were calculated for each row by

$$0.84 \times (\text{standard deviation from Solution 1}) + 25,000. \quad (7.13)$$

Then, the right-hand sides in the linear program for Solution 3 were calculated for each row by

$$0.84 \times (\text{standard deviation from Solution 2}) + 25,000. \quad (7.14)$$

and similarly for Solution 4. This is the sort of simple proportional heuristic used by Heal (1973) and many others. In this case, the procedure seems to converge rather nicely. Global convergence, however, cannot be ensured. Even though the linear program is convex, the post-optimization-calculated statistics are generally not convex with respect to the right-hand sides inserted into the linear program. Without this convexity, iterative convergence cannot be guaranteed (note that Solution 3 in table 7.2 is generally superior to Solution 4). It would thus be risky to automate an iterative search without carefully examining the results. This level of analytical complexity may often be unnecessary anyway. In many planning problems, the sort of rough trial-and-error experimentation demonstrated in table 7.1 generates a combination of confidence and row-total output that is acceptable. The most important point is that decision makers should be presented with solutions that have an associated confidence level identified. This type of information creates a different (and much more realistic) way of viewing the results of an optimization analysis.

TABLE 7.2
An Iterative Approach to Adjusting Right-Hand Sides
with Postoptimization Calculations (all units are in 1000s)

	Right-Hand Side in Linear Program	Row Total Mean	Row Total Standard Deviation	Row Total with 80% Confidence
Solution 1 (cost = 2998.68)				
$t = 1$	35.5	35.5	17.75	20.59
$t = 2$	35.5	35.5	13.94	23.79
$t = 3$	35.5	35.5	12.77	24.78
$t = 4$	35.5	35.5	17.75	20.59
$t = 5$	35.5	35.5	17.21	21.04
$t = 6$	35.5	35.5	13.41	24.23
Solution 2 (cost = 3269.53)				
$t = 1$	39.91	39.91	19.95	23.15
$t = 2$	36.70	36.70	16.70	22.67
$t = 3$	35.72	35.72	13.26	24.58
$t = 4$	39.91	39.91	14.49	27.74
$t = 5$	39.46	39.46	17.77	24.54
$t = 6$	36.27	36.27	17.60	21.49
Solution 3 (cost = 3401.98)				
$t = 1$	41.76	41.76	20.88	24.22
$t = 2$	39.03	39.03	18.98	23.09
$t = 3$	36.14	36.14	15.36	23.24
$t = 4$	37.17	37.17	13.25	26.04
$t = 5$	39.92	39.92	15.19	27.16
$t = 6$	39.78	39.70	14.32	27.75
Solution 4 (cost = 3442.43)				
$t = 1$	42.54	42.54	21.27	24.67
$t = 2$	40.94	40.94	20.43	23.78
$t = 3$	37.90	37.90	17.86	22.90
$t = 4$	36.13	36.13	14.30	24.12
$t = 5$	37.76	37.76	13.37	26.53
$t = 6$	37.01	37.01	13.31	25.83

Row-Total Variance Reduction

Based on the costs reported in tables 7.1 and 7.2, the suboptimality of this approach is considerable (the true minimum cost associated with the output targets and 80 percent confidence is, again, \$2,871,256). Given that this suboptimality occurs because of the possibility of smaller row-total variances, it seems reasonable to attempt to reduce the variances reported in tables 7.1 and 7.2. As discussed in chapter 5, it has long been known that the variance of an investment portfolio can be reduced through diversification of that portfolio as long as the covariances between portfolio components are small relative to their variances (see Markowitz, 1959). Applying this logic to a land allocation problem such as ours would imply diversifying the harvest in any cutting period across analysis areas. The situation in the case example with zero covariances between analysis areas represents something of a worst-case scenario regarding the suboptimality in tables 7.1 and 7.2, but a best-case scenario in terms of potential improvements from land allocation diversification across those analysis areas.

The third column in table 7.3 presents the land allocation for Solution 4 in table 7.2, and the fourth column presents the land allocation for the true optimum that was determined with nonlinear programming. Not surprisingly, the optimal solution is much more diversified than Solution 4 in table 7.2. Indeed, with a constant coefficient of variation across all yield coefficients, the primary difference between these two solutions is that the optimal solution reduces row-total variances through diversification: There are no yields with large means and small standard errors or vice versa. Again, the assumption of zero covariances between analysis areas in the case example clearly enhances the potential for optimality improvements through land allocation diversification.

We tested an approach that combines the postoptimization calculations with a procedure aimed at reducing the row-total variances through diversification of the land allocation. The following (MAXMIN) model is solved with an initial $\tilde{b}$ right-hand side vector as in table 7.2, Solution 1:

Maximize λ

subject to (7.9), (7.10), and

$$\lambda \leq \sum_{k=1}^{n_j} d_{jkt} X_{jk} \qquad \forall\, j \qquad\qquad (7.15)$$

$$\forall\, t,$$

where $d_{jkt} = 1$ if the $a_{jkt} > 0$ and $d_{jkt} = 0$ if the $a_{jkt} = 0$, to obtain λ^*.

TABLE 7.3
Land Allocations from Solution 4 in Table 7.2,
the Optimal Solution, and Solution 4 in Table 7.4

Analysis Area	Harvest Time Periods	Solution 4 in Table 7.2	Optimal Solution	Solution 4 in Table 7.4
1	3	0	269.0	151.5
1	4	637.2	223.7	142.0
1	5	362.8	215.0	294.5
1	6	0	147.4	267.7
2	2	0	363.4	323.1
2	3	836.9	208.8	259.9
2	4	163.1	166.5	139.0
2	5	0	149.0	139.0
2	6	0	112.2	0
2	2 and 6	0	0	139.0
3	1	0	230.9	136.0
3	2	160.8	235.6	308.0
3	3	40.8	125.2	139.0
3	4	0	95.5	139.0
3	5	0	31.2	0
3	6	0	0	0
3	1 and 5	0	113.4	139.0
3	1 and 6	0	168.3	139.0
3	2 and 6	79.8	0	0
4	1	0	202.0	305.0
4	2	0	237.1	139.0
4	3	0	130.5	139.0
4	4	0	99.3	139.0
4	5	0	21.1	0
4	6	0	0	0
4	1 and 5	646.1	141.2	139.0
4	1 and 6	352.5	168.8	139.0
4	2 and 6	1.4	0	0

Then we minimize (7.8) subject to (7.9), (7.10), and

$$\sum_{k=1}^{n_j} d_{jkt} X_{jk} \geq \lambda^* \qquad \forall\, j \qquad (7.16)$$

$$\forall\, t.$$

The $\tilde{\beta}^*$ vector is then calculated with (7.7). If it is unacceptable, a new right-hand side vector $\tilde{b}$ can then be recalculated as in table 7.2, and both models are solved again. Additional adjustments can be made until an acceptable set of 80 percent confidence row totals, calculated with (7.7), are obtained. The idea is to diversify the land allocation across analysis areas (not within analysis areas because of the large covariances) by restricting how much land can be allocated to prescriptions with harvests in a given cutting period in any one analysis area. The first step that maximizes λ actually maximizes the minimum allocation to any one analysis area, which tends to equalize analysis area allocations. Then, the second step is a cost rollover that minimizes cost with the land allocations restricted to be greater than λ^* for all analysis areas in all time periods. It should immediately be pointed out that this procedure may result in slack in any of the constraints in (7.9). If this is undesirable, the λ^* can be arbitrarily decreased in the cost minimization until the slack is eliminated. An alternative (MINMAX) approach would be as follows:

Minimize λ
subject to (7.9), (7.10), and

$$\lambda \geq \sum_{k=1}^{n_j} d_{jkt} X_{jk} \qquad \forall\, j \qquad\qquad (7.17)$$

$$\forall\, t$$

to obtain λ', then minimize (7.8) subject to (7.9), (7.10), and

$$\sum_{k=1}^{n_j} d_{jkt} X_{jk} \leq \lambda' \qquad \forall\, j \qquad\qquad (7.18)$$

$$\forall\, t.$$

In this approach, the first step minimizes the largest allocation to any one analysis area (which again tends to equalize analysis area allocations), and the second step minimizes cost subject to a restriction that no analysis area can be allocated more land than λ'. Both approaches are ad hoc, and both should be tried in a given planning exercise. With the case example, the first approach was more effective, based on cost minimization attainment relative to the 80 percent confidence row totals calculated with (7.7). Other approaches may also be promising.

Table 7.4 presents the results of this type of analysis. As in table 7.2, Solution 1 is the starting point with all right-hand sides in the linear program set to 35,500. The right-hand sides for Solutions 2 and 3 were calculated as just described. Notice that in Solutions 2 and 3, the row total mean exceeds the right-hand side in time period 4. This indicates slack in constraint (7.9). For demonstration purposes, Solution 4 uses the same

TABLE 7.4

An Iterative Approach to Adjusting Right-Hand Sides and Using
a MAXMIN Operator to Diversify Land Allocation
with Postoptimization Calculations (all units are in 1000s)

	Right-Hand Side in Linear Program	Row Total Mean	Row Total Standard Deviation	Row Total with 80% Confidence
Solution 1 (cost = 3118.26, λ* = .1468)				
$t = 1$	35.5	35.5	13.27	24.35
$t = 2$	35.5	35.5	10.57	26.62
$t = 3$	35.5	35.5	9.09	27.87
$t = 4$	35.5	35.5	8.93	28.00
$t = 5$	35.5	35.5	9.49	27.53
$t = 6$	35.5	35.5	9.90	27.19
Solution 2 (cost = 2982.04, λ* = .1488)				
$t = 1$	36.15	36.15	13.36	24.93
$t = 2$	33.88	33.88	10.10	25.39
$t = 3$	32.63	32.63	8.24	25.70
$t = 4$	32.50	34.51	8.76	27.15
$t = 5$	32.97	32.97	9.42	25.05
$t = 6$	33.31	33.31	9.63	25.22
Solution 3 (cost = 2964.87, λ* = .1497)				
$t = 1$	36.22	36.22	13.35	25.00
$t = 2$	33.49	33.49	9.99	25.10
$t = 3$	31.93	31.93	8.07	25.15
$t = 4$	32.36	34.72	8.81	27.31
$t = 5$	32.92	32.92	9.38	25.04
$t = 6$	33.09	33.09	9.50	25.11
Solution 4 (cost = 2920.68, λ* = .139)				
$t = 1$	36.22	36.22	13.66	24.75
$t = 2$	33.49	33.49	9.90	25.17
$t = 3$	31.93	31.93	8.37	24.90
$t = 4$	32.36	32.36	8.21	25.47
$t = 5$	32.92	32.92	9.67	24.79
$t = 6$	33.09	33.09	9.85	24.81

right-hand sides as Solution 3, but with λ^* arbitrarily reduced enough to remove this slack.

In most planning situations, any of solutions 2 through 4 in table 7.4 would probably be acceptable. Solution 4, for example, has 80 percent confidence row totals (calculated with (7.7)) that are very close to the desired 25,000 at a cost that is less than 2 percent higher than the minimum cost in the true optimum. The land allocation for this solution is given in the last column of table 7.3, and is really quite similar to the true optimum solution.

In practice, the true optimum would not be known, so the analyst would never know how much suboptimality persists. As before, however, the confidence levels and right-hand sides *are* accurate (just not necessarily obtained at minimum cost). The MAXMIN and MINMAX procedures can be expected to work well under the case example conditions of a constant coefficient of variation across yield coefficients and zero covariance between yield coefficients across analysis areas. Under these conditions, land allocation diversification is the main distinction between a solution such as Solution 4 in table 7.2 and the optimal solution. If covariances are larger, the suboptimality in Solution 4, table 7.2, would be smaller by the same logic. With highly dissimilar coefficients of variation across yield coefficients, the results are much harder to predict. In many natural resource problems, however, yield standard deviations are roughly proportional to means because the yields are based on growth functions over time such that uncertainty increases with the yield.

The basic premise of this chapter is that it is very important that natural resource analysts begin to present optimization results with some indication of the impact of the uncertainty we face with regard to predicting yields of all outputs. The purpose of the chapter is to suggest some pragmatic approaches, which are immediately applicable to large-scale real-world models. We suggested some postoptimization calculations and demonstrated how these might be performed in a simple case example. The case example assumed stochastic independence between the A-matrix coefficients in different analysis areas, but the procedures do *not* require that assumption. When these covariances are small, the manager may be able to reduce total variability by diversifying the land allocation across analysis areas. We have provided and demonstrated an ad hoc approach to doing this in a linear program. The most important point is, again, that analysts and decision makers recognize uncertainty by reporting results and making decisions that explicitly account for it.

PART III

DYNAMIC MOVEMENT

We now turn to the most fundamental spatial aspect of managed ecosystems: the fact that things move across the landscape over time. Our basic proposition is that many of the ecosystem components that we wish to manage are mobile, so the strategic placement of the management actions (over time) that are targeted at those components is critical. We assume that management actions affect the state of the ecosystem, which in turn affects the mobile components either by affecting their location-specific carrying capacity or by affecting their rate of movement.

Chapters 8 and 9 treat the wildlife habitat placement problem, focusing on habitat connectivity/fragmentation indirectly by modeling wildlife population growth and dispersal. Thus, they address a problem where management activities must be scheduled over time, wildlife habitat changes over time, and different wildlife species respond differently to those habitats. Chapter 8 uses a simple prototype model for demonstration, and chapter 9 presents a real-world case example.

In chapter 10, we address the problem of dynamic pest management with a similar approach. The fact that the mobile component (a pest population) is to be minimized rather than maximized changes the modeling approach significantly. Finally, chapter 11 treats water flow management with a formulation that allows management actions to affect the rate of movement through the watershed. This chapter also describes a unique nested schedule formulation to address the special nature of the water flow management problem. Chapters 10 and 11 are more exploratory and, by necessity, more complex than the other chapters in this section. Unquestionably, other applications for these types of models exist, but the examples provided characterize the types of applications that we can envision at this time.

Methods

A useful method for directly modeling dynamic processes in an optimization problem is to build constraints similar to the equations one would use in a simulation model (see Pielou, 1977; Kitching, 1983; and Chiang, 1984). For example, suppose we wish to model the exponential growth of a population of wildlife species i in some habitat patch h (denoted as population state variable S_{ih}). A dynamic equation for that process at a rate r over some elapsed time t would be

$$S_{iht} = S_{ih0}e^{rt}. \tag{III.1}$$

For many simulation and optimization purposes, we are interested in modeling the growth process over a number of discrete time periods, for which a growth rate is needed. The instantaneous rate of growth for an exponential process (derived from equation (III.1)) is given by the differential equation

$$\frac{dS_{iht}}{dt} = \frac{d(S_{ih0}e^{rt})}{dt} = rS_{ih0}e^{rt} = rS_{iht}. \tag{III.2}$$

Converting from equation (III.2) to a discrete time difference equation for a one-unit change in time, we have

$$S_{ih(t+1)} = S_{iht} + rS_{iht}, \tag{III.3}$$

or

$$S_{ih(t+1)} = (1+r)S_{iht} \qquad t = 0, \ldots, T-1, \tag{III.4}$$
$$S_{ih0} = N_{ih},$$

where we now approximate exponential population growth for species i in patch h over T discrete time periods given an initial population N_{ih}.

Naturally, one drawback to modeling population growth with equation set (III.4) is that the population size is limited only by the initial population size and the number of discrete time periods in the model. Typically, other factors such as amount of habitat (or carrying capacity) also have a limiting effect on the size of the resulting population. Biological models typically capture the effects of carrying capacity with sigmoidal functions such as the following logistic approach:

$$S_{ih(t+1)} \leq (1+r)S_{iht} - \frac{r}{C_{ih}}S_{iht}^2 \qquad t = 0, \ldots, T-1, \qquad \text{(III.5)}$$

$$S_{ih0} = N_{ih},$$

where C_{ih} is the carrying capacity for species i in patch h. We have used an inequality rather than an equality relationship, assuming that limiting factors besides growth rate and carrying capacity might also be included in a complete model. Complex system dynamics pose particular difficulties in optimization models because they often lead to nonconvex programming problems. Because equation set (III.5) is nonconvex, it will probably introduce multiple local optima to the (now) nonlinear programming problem.

In a linear optimization problem, we could model this by replacing equation set (III.5) with

$$S_{ih(t+1)} \leq (1+r)S_{iht} \qquad t = 0, \ldots, T-1, \qquad \text{(III.6)}$$

$$S_{iht} \leq C_{ih} \qquad t = 1, \ldots, T,$$

$$S_{ih0} = N_{ih}.$$

Equation set (III.6) approximates exponential population growth up to carrying capacity, at which point no further growth occurs. It is important to note, however, that equation set (III.6) is well behaved only when the objective function being optimized (whatever it may be) tends to increase population size in all time periods to the point where at least one of the inequalities limits further growth.

A more accurate approximation of (III.5) can be achieved with additional constraints. For example, we could replace (III.6) with

$$S_{ih(t+1)} \leq (1+r)S_{iht} \qquad t = 0, \ldots, T-1, \qquad \text{(III.7)}$$

$$S_{ih(t+1)} \leq S_{iht} + a \qquad t = 0, \ldots, T-1,$$

$$S_{iht} \leq C_{ih} \qquad t = 1, \ldots, T,$$

$$S_{ih0} = N_{ih}.$$

which just adds the additional restriction that after some initial amount of exponential growth (at a point determined by the intersection of the first two inequalities), further growth up to carrying capacity will be linear, based on growth increment a. Although equation set (III.7) provides only an approximation of logistic population growth, it results in a linear program that can readily be solved. Incorporating system dynamics in optimization models is often a matter of making such compromises in order to keep problems tractable.

Our focus in parts I and II was on static optimization. In those parts, we presented static equilibrium and single-time-period models designed to optimize the placement of connected or statistically correlated semi-permanent management actions on a landscape. Dynamic processes contributed to the formulations in only an indirect fashion to the degree that they might affect connectivity or covariance parameters. For example, population growth and dispersal processes are typically among the underlying determinants that establish habitat connectivity probabilities, but the processes themselves were not directly modeled in parts I and II.

In contrast, models based directly on dynamic processes can often be used for optimizing either static or dynamic equilibria. Consider, for example, dynamic equation set (III.8):

$$S_{ih(t+1)} \leq (1+r)S_{iht} + Q_{iht} - R_{iht} \qquad t = 0, \ldots, T-1,$$

$$S_{iht} \leq C_{ih} \qquad t = 1, \ldots, T, \tag{III.8}$$

$$S_{ih0} = N_{ih},$$
$$Q_{ih0} = 0,$$
$$R_{ih0} = 0,$$

where Q_{iht} is the number of animals of species i immigrating into habitat patch h during time period t. Similarly, R_{iht} is the number of emigrants. Note that this is just equation set (III.6) with immigration and emigration variables added, which is the basic device by which we model movement in the forthcoming chapters. At static equilibrium (assuming one exists), any $S_{ih(t+1)}$ must equal the corresponding S_{iht}. Thus, for optimizing static equilibria, equation (III.8) simplifies to

$$rS_{ih} + Q_{ih} - R_{ih} = 0, \tag{III.9}$$
$$S_{ih} \leq C_{ih},$$

where we again use an inequality for the carrying capacity constraint, assuming that the complete model might also include other potential limiting factors.

Given more complex optimization models, including many spatial models, static equilibria become less likely and dynamic equilibria must be considered. Even when static equilibrium exists, we are often just as interested in using a dynamic model to determine a conversion strategy (the management actions required to move from current conditions to equilibrium conditions) as we are in the equilibrium itself. Once a dynamic optimization model is constructed, it is then generally easier to use it for estimating static equilibria than to construct an entirely new static version. We begin with a simple model of wildlife dynamics that includes spatial movement in chapter 8.

Chapter 8

A CELLULAR MODEL OF WILDLIFE
POPULATION GROWTH AND DISPERSAL

In chapter 2 we presented a mixed-integer linear programming approach for optimizing the spatial layout of management actions for wildlife habitat. That chapter was limited to the static case, but accounted for edge effects, habitat size thresholds, and a probabilistic measure of habitat connectivity/fragmentation. This chapter addresses wildlife habitat connectivity/fragmentation (Saunders et al., 1991) by directly modeling wildlife population growth and dispersal; it thus presents a dynamic problem, where management activities must be scheduled over time, wildlife habitat (timber age classes) must be tracked as forest stands age and grow, and different wildlife species respond differently to those habitats. The general problem has a number of similarities to that in chapter 2, but the approach suggested in this chapter is actually quite different. The method is related to earlier reaction-diffusion models used in biodiffusion research by Skellam (1951) and Kierstead and Slobodkin (1953), and in island biogeography research by Allen (1987). Using a continuous reaction-diffusion equation for a single patch with exponential growth to model plankton blooms, Kierstead and Slobodkin were able to demonstrate a critical patch size below which the population perishes. By discretizing the habitat into a number of individual patches, each too small to individually support a persistent population, Allen was

Much of this chapter was adapted from J. Hof, M. Bevers, L. Joyce, and B. Kent, An integer programming approach for spatially and temporally optimizing wildlife populations, *Forest Science* 40(1) (1994): 177–191, with permission from the publisher, the Society of American Foresters.

able to develop and prove several important theorems. These include the existence of a critical number of patches in a linear arrangement of such islands, below which the population perishes. The discrete time approximation used in our spatial optimization model enables application to more general spatial configurations and allows the incorporation of habitat management variables. Discrete time steps also more closely model key life history processes for many species.

Reaction-diffusion models assume that organisms dispersing into unsuitable regions will perish. This mechanism provides a probabilistic basis for the expectation that, after accounting for natality and mortality in an abundant habitat setting through the *r* value (net periodic population growth rate when habitat is not limiting), additional mortality should occur in proportion to the usage of inhospitable surroundings. Our spatial optimization model adopts this assumption, and the amount and arrangement of habitat are dynamically scheduled accordingly.

The Model

We begin, once again, by dividing the land into cells. We then define a set of 0–1 choice variables for each cell, each of which represents a complete, scheduled management prescription. In this case, each prescription defines a time period for harvesting a particular cell, including a no-harvest option. Any harvest resets the timber age class to 0, and changes the habitat carrying capacity for each wildlife species accordingly. Initially, we treat these choice variables discretely (as integer variables), and then we explore relaxation of that assumption. Initial-condition timber age classes are assigned to each cell, as well as initial population numbers for each wildlife species included. The model then chooses one management prescription for each cell.

In order to optimize the spatial layout of management actions that accounts for habitat connectivity over time, some rather specific assumptions regarding wildlife population growth and dispersal are necessary. First, we assume that wildlife populations in any land cell are limited (and therefore determined) by either the carrying capacity of that cell or the combination of growth and dispersal that connects habitat cells from one time period to the next (Kareiva, 1990). We assume that each species has an *r* value that indicates a growth potential per time period in the absence of other limiting factors (in particular, carrying capacity). In addition, we assume that wildlife disperse between one time period and the next in a random fashion. That is, they radiate in all directions (360°) according to some probability density function that relates probability of dispersal to distance. This assumption is consistent with the points raised by Simberloff et al. (1992) as opposed to the movement corridor concept (Harris and Gallagher, 1989; Sessions, 1992). Thus, if carrying capacity is not limiting, the population of each cell in any time period after the first is determined as follows. First, the r value is applied to each cell's population in the previous time period to determine the unconstrained net growth in population between time periods. Then it is assumed that each of these animals has a fixed probability of remaining in the given cell and of dispersing to each other cell between time periods, and that those probabilities decrease with distance according to the probability density function. This density function is defined such that the sum of all these probabilities over some distance in all directions is equal to 1, to account for all animals. Thus the expected value of any cell's population that disperses to any other cell between one time period and the next is the appropriate probability times the previous population of the source cell expanded by the r value. Conversely, the expected value of the population in each cell in any time period can be calculated by summing, across all cells, the products of the associated probabilities and the previous time period's populations expanded by the r value. The same dispersal probability density function is applied to all cells, even if they are close to the boundary of the planning area. It is assumed that any animals that leave the area are no longer included in the planning problem, and that no animals enter the planning problem from external areas. The growth and dispersal process of each species is assumed to be independent of that for other species. It is also important to note that we are assuming wildlife dispersal to be totally random, and that the probabilities are a function only of distance, not population levels.

The basic constraint set for the model is

$$\sum_{k=1}^{q_h} X_{kh} = 1 \qquad \forall\, h, \tag{8.1}$$

$$S_{ih0} = N_{ih} \qquad \forall\, i \tag{8.2}$$
$$\forall\, h$$

$$S_{iht} \le \sum_k a_{ihtk} X_{kh} \qquad \forall\, i \tag{8.3}$$
$$\forall\, h$$
$$t = 1, \ldots, T,$$

$$S_{iht} \leq \sum_n g_{inh}\left[(1+r)S_{in(t-1)}\right] \qquad \forall \, i \qquad \qquad (8.4)$$

$$\forall \, h$$

$$t = 1, \ldots, T,$$

$$F_{it} = \sum_h S_{iht} \qquad \forall \, i \qquad \qquad (8.5)$$

$$\forall \, t,$$

where

i indexes species,

k indexes the management prescription ($k = 1, \ldots, q_h$), where q_h is the number of potential management prescriptions for the hth cell,

h indexes the cells, as does n,

t indexes the time period,

T = the number of time periods,

$X_{kh} \in \{0, 1\}$ = the kth potential management prescription to be considered for the hth cell,

S_{iht} = the expected population of species i in cell h at time period t,

a_{ihtk} = a coefficient set that gives the expected carrying capacity of animal species i in cell h at time period t, if management prescription k is implemented. These numbers are based on the timber age class of cell h in time period t if management prescription k is implemented,

N_{ih} = the initial population numbers for species i in cell h,

g_{inh} = the probability that an animal of species i will disperse from cell n in any time period to cell h in the subsequent time period. This includes a probability for $n = h$, so $\sum_n g_{inh} = 1$ for each combination of h and i,

r_i = the r value population growth rate for species i, and

F_{it} = the total population for species i in time period t.

Equation (8.1) forces the selection of one and only one management prescription for each cell. The management prescriptions are defined with no action in the first time period ($t = 0$), which is used simply to set initial conditions. Equation (8.2) sets the initial ($t = 0$) population numbers for each species, by cell. The S_{iht} (expected population by species, by cell, for $t = 1, \ldots, T$) will be determined by whichever of (8.3) or (8.4) is binding. One or both of them is always binding because of the optimization framework. Constraint set (8.3) limits each cell's population to the carrying capacity of the habitat in that cell, determined by timber age class. Constraint set (8.4) limits each cell's population according to the growth and dispersal from other cells *and itself* in the previous time period. The growth (r_i) and dispersal (g_{inh}) characteristics of each species are reflected by the parameters in constraint set (8.4). Constraint set (8.4) adds up the expected value of the population dispersing from all cells in the previous time period to the given cell in the given time period. It is important to note that whenever (8.3) is binding for a cell, some of the animals assumed to disperse into that cell are lost because of limited carrying capacity. Thus, actual population growth is determined by a combination of potential growth and spatially located limiting carrying capacities. Constraint set (8.5) defines the total population (F_{it}) of each species, in each time period.

Several simple objective functions are explored. For any objective function, it is important to keep in mind that the model will optimize only over the specified time horizon. It is assumed that the basic objective is to manage the area for wildlife. The simplest objective function would thus be to maximize a given species's total population:

Maximize

$$\sum_{t=1}^{T} F_{it} \quad \text{for a given } i. \tag{8.6}$$

Or, a weighted (V_i) sum of species' populations could be maximized:

Maximize

$$\sum_{i} V_i \left(\sum_{t=1}^{T} F_{it} \right). \tag{8.7}$$

The minimum population, over all time periods, for a given species could be maximized (MAXMIN) as follows:

Maximize λ

subject to

$$\lambda \le F_{it} \quad t = 1, \ldots, T \text{ for a given } i. \tag{8.8}$$

Other objective functions are explored in chapter 12.

An Example

The Problem

Let us assume that we have an area that is divided into 25 cells. We include four 10-year time periods: one that defines the initial conditions (time period 0) and three that are subject to management actions. For each cell, we define four potential management prescriptions: cut in time period 1, cut in time period 2, cut in time period 3, and no harvest. All cells are assumed to be 22 time periods (220 years) old initially. Five timber age classes representing important seral stages for wildlife habitat are defined:

ESS = early brush/sapling, age 0–10 years
LSS = late brush/sapling, age 10–20 years
YNG = young forest, age 20–150 years
MAT = mature forest, age 150–250 years
OLD = old growth, age >250 years

Two wildlife species are included with the following carrying capacities, per cell, by timber age class:

	Species 1	Species 2
ESS	4.0	0.0
LSS	3.0	1.0
YNG	2.0	2.0
MAT	1.0	3.0
OLD	0.0	4.0

These are the a_{ihtk} carrying capacity coefficients, which vary by time period and management prescription for each cell. Notice that species 1 prefers younger forest as habitat, whereas species 2 prefers older forest. Thus, alterations in habitat affect each species differently (Merriam et al., 1991). Consistent with these parameters, it is assumed that the initial population of species 1 is 25 (one in each cell), whereas the initial population of species 2 is 75 (three in each cell).

A decadal r value of 3.6 is assumed for both species, such that the unconstrained growth between time periods is a factor of $1 + r = 4.6$. The g_{inh} coefficients for both species are set at .13 for $n = h$, 0.065 for all cells that surround h, 0.022 for all cells that are one-removed from h, and 0.0 for all cells that are farther than that from cell h. Notice that 0.13 $+ (8 \times 0.065) + (16 \times 0.022)$ is (approximately) equal to 1, accounting for all dispersing animals. Also notice that with these parameters, a cell with a totally isolated population will lose 40 percent of its population $[1 - (4.6 \times 0.13)]$ each time period. The two species are assumed to be very similar except for their carrying capacities by timber age class just given. This made model construction and interpretation easier for this example, but different r values and g_{inh} probabilities could be included in the formulation as desired.

Results

Table 8.1 provides the objective function values and wildlife population levels for the solutions in figures 8.1–8.4. Each figure maps the optimal periods of harvest for a particular objective function. In interpreting these solutions, it should be kept in mind that alternative optima are quite likely for this simple example. Also, depending on the objective function, some results may reflect unexploited slack in some variables.

Figure 8.1 presents the spatial layout and timing of harvests when the total (over time periods 1 through 3) population of species 1 is maximized. Species 1 prefers the younger forest stands, so it is not surprising that the entire area is harvested in either periods 1 or 2. A 13-cell diamond (or square) block is harvested in period 1, and then the corners of the planning area are harvested in time period 2. The initial population for species 1 in period 0 is small (25), so the solution harvests the 13-cell block in period 1 to maximize connectivity and take advantage of the desirable habitat created with the harvest. It does not harvest the entire area in period 1, however, because that would cause the population in period 3 to fall off sharply. In period 1, the total carrying capacity of the area is 64 (13 cells with a capacity of 4 and 12 cells with a capacity of 1). The solution (table 8.1) population is 58.8, so the overall limiting factor is dispersal from the small initial population. In period 2, the total carrying

TABLE 8.1
Solution Values for Figures 8.1–8.4

	Figure			
	8.1	8.2	8.3	8.4
Objective function	207.8	300.0	502.68	64.0
Species 1 Population				
$t = 1$	58.80	34.00	50.80	64.00
$t = 2$	87.00	43.00	89.84	64.00
$t = 3$	62.00	42.00	66.00	64.00
Species 2 Population				
$t = 1$	—[a]	66.00	48.00	—[a]
$t = 2$	—[a]	57.00	9.00	—[a]
$t = 3$	—[a]	58.00	32.40	—[a]

[a] Not in the objective function.

2	2	1	2	2
2	1	1	1	2
1	1	1	1	1
2	1	1	1	2
2	2	1	2	2

FIGURE 8.1
Time periods of cell harvest when species 1 population over three time periods is maximized with integer choice variables.

capacity is 87 (13 cells with a capacity of 3 and 12 cells with a capacity of 4). The solution population is also 87, so the overall limiting factor is carrying capacity in period 2. In period 3, the total carrying capacity of the area is 62 (13 cells with a capacity of 2 and 12 cells with a capacity of 3), as is the solution population (table 8.1), so carrying capacity is also limiting in period 3. Thus, the optimal solution finds a balance between obtaining the population increase from period 0 to period 1, where dispersal is limiting, and sustaining that population in periods 2 and 3, where carrying capacity is limiting (as the trees age). The optimal balance is to leave 12 cells for harvest in period 2, but to consolidate the period 1 harvest to get the population up initially, at the expense of dispersing the harvest in period 2. The expense of dispersing the period 2 harvest is not actually very high because all four of the period 2 harvests are adjacent to the period 1 harvest block, so animals are able to populate the new habitat created in period 2 despite its fragmentation into four pieces. This sort of time–space interaction would not be captured in a static analysis of the problem. As pointed out earlier, the selected time horizon is important because the model does not consider conditions beyond that point in time. Forest and wildlife population growth rates would be important considerations in selecting a planning horizon.

When the population of species 2 (over time periods 1 through 3) is maximized, the entire area is left unharvested (which yields 75 animals in periods 1 and 2 and 100 animals in period 3). This is not at all surprising, because this species prefers older timber stands as habitat. The optimal solution is obvious: Let the timber age as much as possible. Note that we do not account for the possibility of natural disturbances that might remove the old timber if it is left alone.

Figure 8.2 presents the spatial layout and timing of harvests when an equally weighted sum of the two species' populations (over periods 1 through 3) is maximized. With this objective function, roughly half of the area is never harvested. The rest is harvested fairly evenly (but with some bias toward later harvests) over time periods 1 through 3. Harvests are spatially clustered much more than nonharvests because the species 1 initial population is so much lower, making the connectivity much more limiting. The species 2 population is much easier to obtain, so that in solution (table 8.1), species 2 fares significantly better than species 1. This shows that the solution reflects not only the objective function coeffi-

3	3	2	0	0
0	2	1	3	3
0	2	2	1	0
3	3	1	0	0
0	0	0	0	0

FIGURE 8.2
Time periods of cell harvest (0 = no harvest) when an equally weighted sum of species 1 and species 2 population over three time periods is maximized with integer choice variables.

2	2	2	2	2
2	1	1	1	2
2	1	1	1	2
2	1	1	1	2
2	2	2	2	2

FIGURE 8.3
Time periods of cell harvest when a
weighted sum (weight on species 1 is twice
that on species 2) of populations over three
time periods is maximized with integer
choice variables.

cients, but also the opportunity costs established by the full set of model
constraints, particularly the initial conditions.

In contrast, figure 8.3 presents the solution when a weighted sum of
the two species' populations is maximized, but with a weight on species
1 that is twice the weight on species 2. With this objective function,
species 1 naturally fares much better. This solution has some obvious
similarity to the one in figure 8.1. This time, a square block of 9 cells is
harvested in time period 1 to increase the species 1 population between
time periods 0 and 1. The total carrying capacity for species 1 in time period 1 is 52 (9 cells with a capacity of 4 and 16 cells with a capacity of
1), whereas the resulting population is 50.8, so dispersal from period 0 is
limiting. The remaining cells, which form the perimeter of the planning
area, are harvested in time period 2. This solution is clearly geared to-
wards species 1 (see table 8.1), but not as much as the solution in figure
8.1. As in figure 8.1, however, the solution spatially clusters the management action that favors the species whose limiting factor is dispersal
from a previous time period. When a weighted sum of the two species'
populations was maximized with a weight on species 2 that was twice

2	1	1	1	1
2	1	1	1	3
1	1	1	1	3
3	1	1	1	3
3	1	1	1	3

FIGURE 8.4
Time periods of cell harvest when the minimum species 1 population over three time
periods is maximized with integer choice
variables.

that on species 1, the solution was again to simply leave the area totally
unharvested.

Figure 8.4 presents the solution when, across time periods 1 through
3, the minimum species 1 population is maximized. The solution for this
figure is not necessarily optimal because the enumeration (implicit or explicit) of all possible integer solutions could not be completed within
reasonable computing time. Based on the bound from the continuous
linear program used in the solution algorithm, it is within 5 percent of
optimality in terms of objective function attainment. The solution in figure 8.4 is also rather similar to those in figures 8.1 and 8.3. Because of
the MAXMIN operator, however, an even species 1 population across
time periods 1 through 3 was obtained (table 8.1). This required significantly more harvesting in time periods 1 and 3, and less in period 2. In
particular, 17 cells were harvested in time period 1 (again clustered together to maximize connectivity) to bring its population up to its optimal
level. Time period 1 is obviously the most limiting period with the
MAXMIN operator because of the limiting dispersal from the low population of species 1 in the first time period. Also, more harvesting in time
period 3 was optimal, along with less in time period 2, to maintain the

population established in time period 1. The MAXMIN operator clearly creates a schedule different from the simple maximization of total species 1 population. When the MAXMIN operator was applied to the species 2 population, not surprisingly, the entire area was left unharvested.

Even with the simple dimensions of the case example, a fairly large mixed-integer program was necessary. We automated the generation of the model; otherwise, its construction would have been quite onerous. Solution times were extremely unpredictable. Solved initially as linear programs, some of the models resulted naturally in integer or near-integer solutions, which provided very rapid optimization. At the other extreme, the MAXMIN solution in figure 8.4 is not necessarily optimal because we were unable to completely enumerate (implicitly or explicitly) all possible integer solutions. Because each branch in the solution provides a bound on optimality, however, those solutions are within a known percentage of being optimal. Even though the attempt at complete enumeration was allowed to run for over 200 hours on a 486/33 machine, the MAXMIN solution reported was obtained within the first hour. In practice, satisficing solutions to the integer models within a given tolerance of optimality are probably a realistic expectation, especially with the MAXMIN operator.

Continuous Choice Variables

In many cases, it is tenable to treat the X_{kh} choice variables as continuous variables bounded by

$$0 \leq X_{kh} \leq 1 \qquad \forall k \qquad (8.9)$$
$$\forall h.$$

Relaxing the integer constraint requires an assumption that all responses to management actions on each cell are proportional to the amount of the cell treated. With homogeneous cells, this implies an assumption that the a_{ihtk} coefficients apply (linearly) to any proportion of the cells, not just the total cell. If the carrying capacity is actually a nonlinear function of habitat area, the a_{ihtk} coefficients might well be tenable for the (fixed-size) integer choice variables and not be tenable for the (variable-sized) continuous choice variables. With nonhomogeneous cells, relaxing the integer constraint typically implies an additional assumption that all management actions within a cell are applied proportionally to each type of land within the cell. Suppose a 100-acre cell (h) is made up of 25 acres of young, 50 acres of mature, and 25 acres of old timber, and X_{1h} (harvest in period 1) solves at .4 whereas X_{2h} (harvest in period 2) solves at .6. This implies that 10 acres of young, 20 acres of mature, and 10 acres of old timber are treated with scheduled prescription X_{1h}; and 15 acres of young, 30 acres of mature, and 15 acres of old timber are treated with scheduled prescription X_{2h}. It must then still be tenable to assume that the a_{ihtk} coefficients in equation (8.3) are constants for any portion of their cells, such that each portion is defined with the proportional representation of land types initially present in those cells.

In addition, relaxing the integer constraint requires an assumption that the location of habitat within a cell is unimportant. All the spatial data (especially the g_{inh} coefficients) would have to reference a single point for each cell (such as its centroid), and that point would have to serve as the locator for all activity in each cell. These assumptions are stronger than in the integer variable case, but the advantages of solvability and added choice variable flexibility in the continuous-variable approach are quite attractive in many cases.

Results

Figures 8.5, 8.6, 8.7, and 8.8 present solutions to models identical to those presented in figures 8.1, 8.2, 8.3, and 8.4, respectively, except that the models for figures 8.5–8.8 use continuous choice variables. Table 8.2 presents the objective function values and wildlife population levels for the solutions in figures 8.5–8.8. The solutions in figures 8.5 and 8.7 are almost identical, and are something of a compromise between the solutions in figures 8.1 and 8.3. The same basic pattern, however, is clearly evident in figures 8.1, 8.3, 8.5, and 8.7: A square block is harvested in time period 1, with the outer perimeter being harvested in later time periods. In comparing table 8.1 with table 8.2, the solutions in figures 8.5 and 8.7 gain modest improvements in objective function attainment over their table 8.1 counterparts because of the increased choice variable flexibility. It is also noteworthy that the schedule of species populations is significantly different with the continuous-variable solutions for figures 8.5 and 8.7. The possibility of alternative optima should be remembered, but it appears that the continuous choice variables allow a solution with more early harvesting to generate early species 1 population, with a lower opportunity cost in terms of species 1 population in periods 2 and 3. We are able to cut part of the perimeter in period 1 and still retain some of it for harvesting in periods 2 and 3 (to generate species 1 habitat in those periods). The gain in the objective function over all three time

TABLE 8.2
Solution Values for Figures 8.5–8.8

	Figure			
	8.5	8.6	8.7	8.8
Objective function	217.13	300.00	516.88	67.19
Species 1 Population				
$t = 1$	76.40	33.30	76.40	67.19
$t = 2$	82.87	45.31	82.48	67.19
$t = 3$	57.87	39.36	57.99	67.19
Species 2 Population				
$t = 1$	—[a]	66.70	23.60	—[a]
$t = 2$	—[a]	54.69	17.52	—[a]
$t = 3$	—[a]	60.64	42.00	—[a]

[a] Not in the objective function.

1: .33 2: .67	1: .57 2: .43	1: .66 2: .33	1: .57 2: .43	1: .33 2: .67
1: .57 2: .43	1: .9 2: .1	1: 1	1: .9 2: .1	1: .57 2: .43
1: .67 2: .33	1: 1	1: 1	1: 1	1: .67 2: .33
1: .57 2: .43	1: .9 2: .1	1: 1	1: .9 2: .1	1: .56 2: .43
1: .33 2: .67	1: .57 2: .43	1: .66 2: .33	1: .57 2: .43	1: .33 2: .67

FIGURE 8.5
Time periods and proportions of cell harvest when species 1 population over three time periods is maximized with continuous choice variables.

periods is small, but increased scheduling flexibility is certainly created by the continuous choice variables. The advantage in solvability is even more apparent: As expected, all the continuous-variable models solve almost immediately.

In comparing the solution in figure 8.6 with that in figure 8.2, we once again see a great deal of similarity. With a nontrivial emphasis on species 2, roughly half the area is not harvested and the rest of the area is harvested across time periods 1, 2, and 3. In figure 8.6, a bit more of the area is harvested, with a bit less late harvesting. And, in both figures, the harvests are somewhat more clustered than the unharvested cells (to connect the species 1 population that must be generated through harvesting). In comparing tables 8.1 and 8.2, the objective function attainment for these two solutions is identical, and the schedules of species populations are quite similar.

In comparing the solution in figure 8.8 with that in figure 8.4, we see a very different spatial layout but a common emphasis on harvesting in time periods 1 and 3 so as to achieve the even species 1 population across time periods 1, 2, and 3. All but two of the cells are at least par-

2: .44 3: .56	2: .77 3: .23	0	0	0
1: .23	0	1: .33 3: .67	1: .34 3: .66	3: 1
0	2: 1	0	0	2: .83
1: .47 3: .53	1: .73 3: .27	2: 1	0	0
1: .30	1: .09	0	1: .27 2: .63 3: .10	2: .26 3: .74

FIGURE 8.6
Time periods and proportions of cell harvest (0 = no harvest) when an equally weighted sum of species 1 and species 2 population over three time periods is maximized with continuous choice variables.

1: .33 2: .64 3: .03	1: .57 2: .43	1: .67 2: .33	1: .57 2: .43	1: .33 2: .64 3: .03
1: .57 2: .43	1: .9 2: .1	1: 1.0	1: .9 2: .1	1: .57 2: .43
1: .67 2: .33	1: 1.0	1: 1.0	1: 1.0	1: .67 2: .33
1: .57 2: .43	1: .9 2: .1	1: 1.0	1: .9 2: .1	1: .57 2: .43
1: .33 2: .64 3: .03	1: .57 2: .43	1: .67 2: .33	1: .57 2: .43	1: .33 2: .64 3: .03

FIGURE 8.7
Time periods and proportions of cell harvest when a weighted sum (weight on species 1 is twice that on species 2) of populations over three time periods is maximized with continuous choice variables.

1: .32 3: .68	1: .55 3: .45	1: .66 3: .34	1: .57 2: .43	1: .33 2: .67
1: .54 2: .46	1: .89 2: .11	1: 1.0	1: .87 3: .13	1: .54 3: .46
1: .65 3: .35	1: 1.0	1: .06 3: .94	1: .95 3: .05	1: .58 2: .42
1: .56 2: .44	3: 1.0	1: 1.0	1: .81 3: .19	1: .39 2: .61
1: .33 3: .67	1: .57 2: .43	1: .64 2: .36	3: 1.0	1: .25 2: .75

FIGURE 8.8
Time periods of cell harvest when the minimum species 1 population over three time periods is maximized with continuous choice variables.

tially harvested in time period 1 in figure 8.8, followed by scattered cutting in time periods 2 and 3. This is clearly in the spirit of the solution in figure 8.4, but the continuous variables allow a different spatial approach. The resulting objective function attainment (table 8.2 versus table 8.1) is the most improved of the four choice-variable comparisons, but is still modest at about 5 percent. The increased flexibility in laying out harvests over time and space in figure 8.8 is clear, however. The difference in solvability is most pronounced in this comparison, because we could not even ensure optimality in figure 8.4 (the solution in figure 8.8 is actually the bound on that integer solution, as previously reported). With more complicated objective functions such as the MAXMIN operator, the advantages of the continuous choice variables are obvious.

Discussion

The tenability of the basic approaches discussed here probably rests on the rather rigid assumptions made regarding wildlife population growth and dispersal over time. These phenomena are not well understood and certainly vary from species to species (Wiens, 1992; Saunders et al., 1991). The independence assumptions between species are also quite strong. Expansion of the model beyond the simple demonstrative example is fairly straightforward. Additional wildlife species could be added with additional constraints, but without the need for additional habitat choice variables. Additional management options (or prescriptions), including different kinds of vegetative manipulation or multiple management actions over time, would require additional choice variables, but in proportion to the number of those options. Likewise, additional numbers of cells (spatial resolution) could be added with a proportional increase in the number of choice variables. Additional time periods could also be added with additional choice variables. So long as the number of management options per time period is fixed, this increase in the number of choice variables would be in proportion to the number of time periods. The number of habitat classes could be increased with no impact on model size. In chapter 9, we present a real-world application of the continuous-variable model formulation, with a number of these expansions included.

Chapter 9

THE BLACK-FOOTED FERRET: A CASE STUDY

In this chapter, we present a wildlife habitat spatial optimization case study for recovery of an endangered species using the continuous habitat management choice variables described in chapter 8. Through this large-scale, real-world model a number of results arise that could not be seen as readily with our smaller example.

Early in 1987 the black-footed ferret (*Mustela nigripes*) became one of the world's most endangered mammals when the last known free-ranging member of the species was taken into captivity (Thorne and Belitsky, 1989). Under the care of the Wyoming Game and Fish Department, six of the eighteen surviving ferrets were successfully bred in captivity (Clark, 1989). This set the stage for a national recovery program of releasing captive-bred ferrets back into the wild.

Historically, the black-footed ferret ranged sympatrically with prairie dogs (*Cynomys* spp.) across much of North America (Anderson et al., 1986). Available evidence strongly supports the conclusion by Henderson et al. (1969) that black-footed ferrets are a stenecious species (having narrow habitat requirements), living principally in prairie dog burrows and depending primarily on prairie dogs for prey (Linder et al., 1972). Demise of the species in the wild has been attributed to loss and fragmentation of habitat (prairie dog colonies) due to extensive prairie

This chapter was adapted from M. Bevers, J. Hof, D. W. Uresk, and G. L. Schenbeck, Spatial optimization of prairie dog colonies for black-footed ferret recovery, *Operations Research* 45(4) (1997): 495–507, with permission from the publisher, the Institute for Operations Research and the Management Sciences (INFORMS).

dog eradication (rodenticide) programs, sylvatic plague, and changes in land use, combined with susceptibility to canine distemper (U.S. Fish and Wildlife Service, 1994). As Seal (1989) points out, it is now difficult to find suitable ferret habitat complexes ("groups of prairie dog colonies in close proximity," Biggins et al., 1993) of 3000 to 15,000 ha, even though prairie dogs once occupied 40 million ha of land.

The first release of captive-bred black-footed ferrets into the wild occurred in 1991 in Shirley Basin, Wyoming. Two additional reintroduction areas were added to the recovery program in 1994, including the site of this study, centered in Badlands National Park, South Dakota. These ferret release sites were selected on the basis of habitat suitability and other biological and sociopolitical factors. Prairie dog population management within these sites will be a critical component in the success or failure of ferret reintroductions. Rodenticides are actively used in the northern Great Plains, and have greatly reduced prairie dog populations (Roemer and Forrest, 1996). Black-footed ferret recovery at the Badlands reintroduction site will probably be affected by the location and timing of rodenticide treatments in the area.

The spatial arrangement of prairie dog colonies in a complex is recognized as having important ramifications for the number of black-footed ferrets that can be supported (Minta and Clark, 1989). Houston et al. (1986) and Miller et al. (1988) have used spatial measures such as mean intercolony distance and colony size frequency distribution in estimating ferret habitat suitability, but Biggins et al. (1993) note a number of troubling quantitative difficulties with such approaches. Consequently, in the Biggins et al. procedure the effects of spatial colony arrangement within colony complexes are assessed qualitatively. The purpose of this case study (see Bevers et al., 1997) is to present the use of a more rigorous quantitative approach, based on the theoretical concepts discussed in chapter 8. The resulting model is used to explore habitat evaluation and management as a spatial efficiency problem on the federally managed lands of the Buffalo Gap National Grassland adjacent to the Badlands National Park ferret release area.

Spatial Optimization Model

Within a reintroduction area, black-footed ferret habitat comprises a complex of active and potential prairie dog colonies (patches) forming distinct habitat islands on the landscape. We impose a grid of habitat management cells on this complex, and use discrete time periods so that all potential spatial configurations can be considered (within the resolution of our grid). Rodenticide treatments, which have a negative effect on black-footed ferrets by reducing prairie dog numbers, are the principal habitat management tool to be considered. Particular ferret habitat layouts are achieved over time by the prairie dog populations that result from the rodenticide treatment/nontreatment schedules applied to each cell across the landscape, on the premise that prairie dog populations will recover rapidly in areas left untreated. Annual time periods are used to closely model key ferret life history processes.

Adult black-footed ferret populations expected in each cell in any year are then limited by the carrying capacity for that cell, or the ability of ferrets from nearby cells to successfully reproduce and disperse there, or both. Additional decision variables are used to determine the timing and location for captive-bred ferret releases into the area. The solution of the model indicates a complex of prairie dog colony populations, over time, that supports a dynamic black-footed ferret population. Our spatial optimization model is as follows:

Maximize

$$F_T, \tag{9.1}$$

subject to

$$F_t = \sum_i S_{it} \qquad t = 1, \ldots, T, \tag{9.2}$$

$$S_{i0} = N_i \qquad \forall\, i, \tag{9.3}$$

$$S_{it} \le R_{it} + \sum_j g_{ij}(1 + r_j)S_{j(t-1)} \qquad \forall\, i \qquad t = 1, \ldots, T \tag{9.4}$$

$$\sum_i g_{ij} \le 1 \qquad \forall\, j,$$

$$\sum_i R_{it} \le b_t \qquad t = 1, \ldots, T, \tag{9.5}$$

$$S_{it} \le \sum_{h=1}^{m_i} \sum_{k=1}^{n_{ih}} c_{ihkt} X_{ihk} \qquad \forall\, i \qquad t = 1, \ldots, T, \tag{9.6}$$

$$\sum_{k=1}^{n_{ih}} X_{ihk} = A_{ih} \qquad \forall\, i, h, \tag{9.7}$$

$$\sum_i \sum_{h=1}^{m_i} \sum_{k=1}^{n_{ih}} c_{ihktp} X_{ihk} \le C_{pt} \qquad \forall\, p \qquad t - 1, \ldots, T \tag{9.8}$$

$$X_{ihk}, S_{it}, R_{it} \ge 0 \qquad \forall\, i, h, k, t.$$

Indices:

t indexes annual time periods from year 0 up to T that begin when the young emerge from the den, typically around May or June of each year,

i and j both index the entire set of habitat management cells in the study area,

h indexes the set of different initial habitat condition classes available in each cell,

k indexes the set of rodenticide treatment schedules being considered for each habitat condition class in each cell, and

p indexes the set of policy constraints (if any) on selection of habitat management variables.

Decision variables:

X_{ihk} = the amount of area (in hectares) in cell i and initial habitat condition class h allocated to the kth multiyear habitat management schedule of rodenticide treatment or nontreatment being considered for that cell and

R_{it} = the number of captive-bred ferrets released in cell i during year t that are expected to survive adaptation to life in the wild.

Response variables:

S_{it} = the expected adult ferret population (including yearlings) in cell i at the beginning of year t, plus R_{it} and

F_t = the expected adult ferret population for the entire complex in year t.

Parameters:

N_i = the actual or estimated initial number of adult ferrets in cell i,

g_{ij} = the proportion of surviving adult and juvenile ferrets from den areas in cell j in year $t - 1$ expected to disperse and become adult ferrets (subject to habitat availability) in cell i den areas at the beginning of year t,

r_j = an r value for ferrets in cell j reflecting the maximum expected annual net population growth rate, that is, the growth rate when habitat is in no way a limiting factor,

b_t = an upper bound on the total number of captive-bred ferrets released during year t expected to survive adaptation to life in the wild,

c_{ihkt} = the expected adult black-footed ferret carrying capacity for cell i and condition class h in year t per hectare allocated to X_{ihk},

m_i = the number of initial habitat condition classes in cell i,

n_{ih} = the number of habitat management schedules being considered for habitat condition class h in cell i,

A_{ih} = the total prairie dog colony area (in hectares) of initial habitat condition class h in cell i,

$c_{ihktp} = c_{ihkt}$ if X_{ihk} could contribute to policy constraint p in time period t, and is 0 otherwise, and

C_{pt} = the amount of total expected black-footed ferret carrying capacity allowed in time period t under policy p from the relevant subset of X_{ihk} habitat management variables.

This model is based on the general ideas in chapter 8, but reflects the specific ferret-release problem addressed, which includes a nonhomogeneous landscape with patches of current and potential habitat. Equations (9.1)–(9.5) define a discrete-time reaction-diffusion system (in this case, population growth and dispersal) for evaluating population persistence within a complex of habitat cells. Equations (9.6) and (9.7) account for these habitat dynamics by imposing black-footed ferret carrying capacity constraints as a function of the selected habitat cell management (rodenticide treatment) schedules. Consequently, the expected ferret population in any cell in a given year (S_{it}) is determined by either equation (9.4) or (9.6), whichever is limiting.

The linear dispersal model (equation 9.4) requires an assumption of purely random diffusion, which can be considered realistic only as a first-level approximation for a highly developed species such as the black-footed ferret. More realistic ferret dispersal patterns, if they were known, might exhibit biased diffusion, such as movement in response to overcrowding (Gurney and Nisbet, 1975) or somewhat selective movement based on acquired knowledge of active prairie dog colony locations or surrounding terrain. In general, biased diffusion enhances the persistence of populations (Allen, 1983). Thus, our model provides an estimate that can be viewed as a lower bound on the size of the expected population.

In some cases, exponential population growth within cells may be unreasonably optimistic. For more conservative growth rates, sigmoidal population growth could be approximated in a linear model by replacing equation (9.4) with

$$Q_{jt} \leq (1 + r_j)S_{j(t-1)} \qquad \forall\, j \qquad (9.9)$$
$$t = 1, \ldots, T,$$
$$Q_{jt} \leq S_{j(t-1)} + a_j \qquad \forall\, j$$
$$t = 1, \ldots, T,$$

$$S_{it} \le \sum_{j} g_{ij} Q_{jt} \qquad \forall i$$

$$t = 1, \ldots, T,$$

where an accessory variable for cell population in each year, Q_{jt}, is limited to either exponential (r_j) or incremental (a_j) growth up to carrying capacity, depending on the magnitude of $S_{j(t-1)}$. Note that we have identified the r_j and a_j growth rates by cell. In some cases, expected net reproduction may be different from one cell to another even with abundant habitat. For example, one cell may lie closer to terrain frequented by predators than another cell. Random dispersal and carrying capacity limits combine to provide a sigmoidal growth dynamic for expected black-footed ferret population over the entire prairie dog colony complex (F_t) when a persistent population is possible, even though cellular population growth is exponential in our model.

Dispersal in equation (9.4) assumes that ferrets within the complex originate from reintroduced animals. By limiting the total annual release, as in equation (9.5), the spatial optimization model is used to help identify preferred ferret release locations. If release locations are predetermined, the annual cellular release variables (R_{it}) could be individually constrained instead of the annual sum. We also assume that released ferrets will disperse and reproduce similarly to indigenous ferrets once they are acclimated to life in the wild. Additional mortality during the establishment period is accounted for by adjustments to the upper bounds on releases (b_t). If released ferrets were known to initially disperse or reproduce differently, equation (9.4) could be replaced with

$$S_{it} \le \sum_{j} \left[g_{ij}(1 + r_j)S_{j(t-1)} + g_{ij}^{R}(1 + r_j^{R})R_{j(t-1)} \right] \qquad \forall i \quad (9.10)$$

$$t = 1, \ldots, T$$

$$\sum_{i} R_{i0} = 0.$$

This would defer accounting for releases in the S_{it} and F_t variables for 1 year.

The linear programming approach used here leaves the specific locations of habitat resulting from different treatment schedules within a cell unresolved. The choice of cell size controls the spatial resolution of the model. If very large cells were used, rodenticide treatment locations would be less well specified, which could reduce the accuracy of the g_{ij} dispersal coefficient estimates. The size of these errors can be controlled by using smaller land units (cells) in the model, in a manner typical of numerical approximations. As smaller and smaller cells are used, the dispersal estimation error approaches 0 but model size increases. In practice, some compromise is required to address large-scale problems. Dispersal estimation errors are reasonably small if we define habitat management cell sizes that are small relative to ferret dispersal ranges.

In a case such as black-footed ferret reintroductions, where initial conditions differ greatly from potential persistent population levels, habitat conversion strategies based on prairie dog population management decisions in the first several years can be as important as long-term management strategies. The ability to address conversion strategies is one important advantage of a dynamic model. The relative weight placed on short-term versus long-term management depends primarily on the choice of objective function (see chapter 13). Although the objective function expressed in equation (9.1) is suitable for estimating persistent expected ferret population levels, another objective function was also used in the case study to place more emphasis on early ferret establishment. We replaced equation (9.1) with the following:

Maximize

$$\sum_t F_t \tag{9.11}$$

for all analyses except those focused on long-term persistence. The spatial optimization model used throughout the case study combines either equation (9.1) or equation (9.11) as an objective function, with equations (9.2)–(9.8) as constraints.

Ferret Reintroduction in South Dakota

Between September 19 and November 14, 1994, 36 captive-bred black-footed ferrets were released near the center of the Sage Creek Wilderness in Badlands National Park (McDonald, 1995). These ferrets made up the first of approximately five annual fall releases planned for the purpose of reestablishing ferrets in the Park and the surrounding Buffalo Gap National Grassland (Plumb et al., 1994).

We applied the spatial optimization model to approximately 1575 km^2 of land in and around the South Dakota reintroduction area, assuming that prairie dog population controls preclude ferret recovery outside this study area. Federally managed lands with active or readily recoverable black-tailed prairie dog (*C. ludovicianus*) colonies suitable for ferret habitat over the next 10–15 years are fragmented and occupy less than one-tenth of the study area. Badlands National Park contains approximately 24 km² of prairie dog colonies, primarily in the rugged Sage Creek Wilderness. The acreage of prairie dog colonies in the Park is not expected to change significantly over the next 10–15 years. Under current management plans, Buffalo Gap National Grassland supports approximately 21 km² of predominantly active prairie dog colonies reserved from rodenticide use adjacent to Badlands National Park. We refer to these as current colonies throughout this chapter. The Grassland contains an additional 99 km² of predominantly inactive prairie dog colonies in the study area that have been treated with rodenticide in past years. We refer to these as potential colonies throughout this chapter.

Spatial Definition

We selected U.S. Public Land Survey sections as our basic land mapping unit for the model, and assumed for dispersal distance calculations that each of the 608 survey sections was a square enclosing 2.59 km² of land. The number of hectares of existing and potential prairie dog colonies within each section was estimated from color infrared aerial photography taken in 1983 and 1993 at a scale of 1:16,000 using methods described by Schenbeck and Myhre (1986) and Uresk and Schenbeck (1987). The 1983 prairie dog colony distribution was used as an estimator of potential colony distribution because this was the period when recorded prairie dog populations were greatest. Habitat areas within each survey section were classified as either National Park Service–administered lands, USDA Forest Service–administered lands presently subject to prairie dog population controls (potential), or USDA Forest Service–administered lands presently reserved from prairie dog population controls (current). Lands under private ownership were not included in the model.

Ferret Dispersal

Although few observations of ferret movements are available, distances of 2–3 km were common for both nightly movements and annual inter-colony movements (primarily by juveniles in late summer or early fall) of wild-born ferrets at Meeteetse, Wyoming (Forrest et al., 1985; Biggins et al., 1986; Richardson et al., 1987). Statistics reported by Oakleaf et al. (1992, 1993) from captive-bred ferret releases at the Shirley Basin prairie dog colony complex in Wyoming roughly suggest an exponential distribution of dispersal distances. Dispersal appears to have been equally likely in all directions. Eight of the ferrets released in Badlands National Park in 1994 were observed to disperse with a mean distance of 3.7 km and a maximum distance of 11.8 km (standard deviation = 4.2 km) over about a 30-day period. It is not known to what degree differences between these observations result from differences between captive-bred and wild-born ferrets, differences between prairie dog colony complexes, or from other causes.

For this study, we assumed that all ferrets will disperse annually according to an exponential distance distribution with a mean of 3.7 km in uniformly random directions over a radius of about 14 km. Dispersal coefficients (g_{ij}) were then estimated by numerical approximation of the integral of this bivariate dispersal distribution over distances and angles defined by the boundaries of each destination (i) section relative to the center of each source (j) section. Differences in dispersal behavior between sexes have been ignored, as have differences between released ferrets and wild-born ferrets. The effects of rugged topography in the Badlands could be taken into account in the pairwise estimation of dispersal coefficients, but these effects are unknown.

Net Population Growth Rate

Wild ferrets have not been studied under conditions of unlimited habitat. Consequently, field observations of appropriate r values (r_j) for this model are lacking. Instead, an r value was estimated by simulating unlimited population growth using mean natality and mortality rates and initial conditions similar to those assumed by Harris et al. (1989) in their research on black-footed ferret extinction probabilities. After 12 simulation years the population ratios and growth rates stabilized with an r value (net annual population growth rate) of 0.8175. This r value was used for all cells in the model.

Ferret Releases

Based on past experiences (Oakleaf et al., 1992, 1993), approximately 80 percent of released captive-bred black-footed ferrets are expected to die during their first 30 days in the wild. Half of the remaining ferrets are expected to perish during their first winter. Of the 36 ferrets released in Badlands National Park in 1994, only 8 were known to survive the first 30 days. Taking into account likely winter mortality, we assigned an expected population value of 0.5 adult ferrets at each of the eight surviving ferret locations as initial conditions (N_{i0}) in the model (with zeroes as-

signed elsewhere). We expected 40 more ferrets to be released in the fall
of each of the following four years. Assuming that 4 ferrets from each
release would survive to reproduce, we set b_1 through b_4 equal to 4.0
(with zeroes assigned for all additional years).

Ferret Carrying Capacity

Although black-tailed prairie dog densities within colonies typically decline over extended periods of time (Cincotta, 1985; Hoogland et al.,
1988), existing colonies in the Badlands area are expected to remain well
populated for the next 10–15 years. Based on the observations reported
by Hillman et al. (1979), ferret carrying capacities on existing or fully
recovered prairie dog colonies were set at slightly over 0.05 ferrets/
hectare. The map in figure 9.1 shows the spatial arrangement of current
plus potential ferret habitat by survey section in the study area at maximum model carrying capacity. Based on studies by Knowles (1985),
Cincotta et al. (1987), and Apa et al. (1990), we estimated that complete
prairie dog population recovery in recently treated colonies would require an average of four breeding seasons. We set ferret carrying capac-

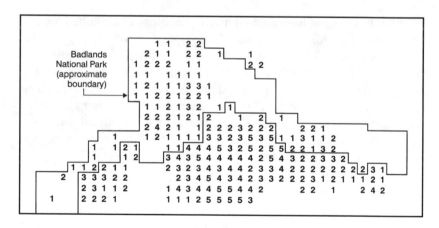

Map code 1 = (0, 1] Black-footed ferrets
2 = (1, 4]
3 = (4, 7]
4 = (7, 10]
5 = (10, 13.66)

FIGURE 9.1

Short-term (10- to 15-year) potential black-footed ferret carrying capacities
on federally managed lands within the study area.

ity accordingly at one-eighth of full capacity in the first year following use of rodenticide, at one-fourth of full capacity in the second year, at one-half of full capacity in the third year, and at full capacity thereafter. This rate of recovery may require special management actions, such as intensive livestock grazing, to aid the spread of prairie dogs (Uresk et al., 1981; Cincotta et al., 1988). We also assumed that all potential habitat areas in the model could begin recovery in any chosen year.

Model Results

The model was solved using equation (9.11) as the objective function, with equation (9.8) set to various right-hand side levels to restrict additions of ferret carrying capacity from potential prairie dog colonies on the Grassland. A 25-year planning horizon ($T = 25$) was used to allow enough time for expected population levels to stabilize, but care must be taken not to overinterpret results beyond 10–15 years. The graph in figure 9.2 shows the results of allowing no additional habitat, and with 20, 40, 60, 80 and 100 percent additions of the potential Grassland carrying capacity. In all cases, sigmoidal expected population growth curves resulted. As we would expect, the graph shows diminishing marginal returns as more carrying capacity is added because the most spatially efficient habitat areas are included first. It is interesting to note that each curve levels off substantially below total carrying capacity. For example, when all Grassland habitat areas are allocated to prairie dog colonies the expected population of ferrets rises to only about 85 percent of the summed habitat capacity of more than 757 ferrets. This suggests that simply totaling available habitat tends to overestimate actual carrying capacity because spatial effects have not been taken into account.

Preferred habitat areas change over time, as shown in figure 9.3 by maps of the habitat allocated in different years under the alternative that adds 20 percent of the potential Grassland carrying capacity for ferrets. The 20 percent constraint for this alternative is binding on the solution from year 7 on. Before that year, the expected ferret population is still small enough that habitat is not limiting. Figure 9.3a shows the habitat allocations for year 7. Survey sections shaded in gray are predominately potential Grassland habitat areas, where management choices (schedules of rodenticide treatment) were allowed. In figure 9.3a, a small amount of habitat is allocated to all but one survey section with potential habitat as the fledgling expected ferret population is rapidly expanding throughout the area (compare with figure 9.1). The expected population in year 7 is

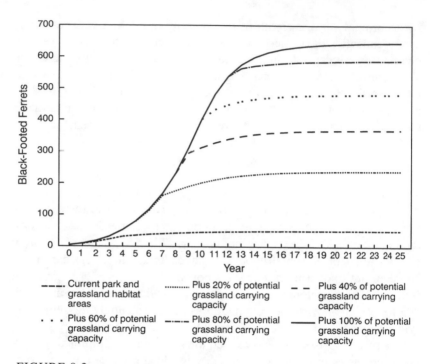

FIGURE 9.2
Expected black-footed ferret populations under the present management strategy and five alternative strategies.

shown in figure 9.4b, along with the corresponding selected ferret release locations shown in figure 9.4a.

By year 15, the expected ferret population under this alternative has largely leveled off at more than 230 adult ferrets, and the preferred habitat allocations have shifted to more concentrated areas around the Park and current Grassland colonies (figure 9.3b). Dashes on this map indicate survey sections that have some habitat allocated in year 7 but no longer have any habitat allocated by year 15. The corresponding expected ferret population is shown in figure 9.4c.

We examined the habitat allocations for this alternative near equilibrium conditions by bounding all ferret release variables to 0, unbounding all initial population variables, and using equation (9.1) as the objective function. We also allowed the model to schedule treatments for Grassland colonies currently reserved from rodenticide use in place of potential colony treatments to allow greater freedom for spatial optimization. The resulting allocations shown in figure 9.3c support an expected population that levels off at a little more than 300 adult ferrets. Note that the

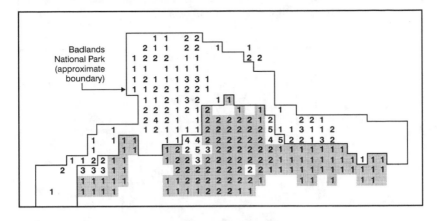

Map code **1** = (0, 1]   Black-footed ferrets
           2 = (1, 4]
           3 = (4, 7]
           4 = (7, 10]
           5 = (10, 13.66)
           ▨ = National Grassland areas with
               management choices available

FIGURE 9.3a
Habitat allocations expressed as black-footed ferret carrying capacities under the +20% alternative in year 7.

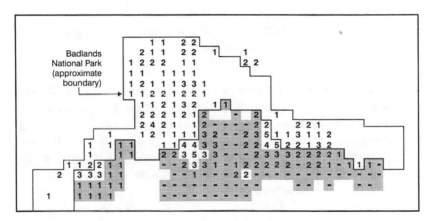

Map code **1** = (0, 1]   Black-footed ferrets
           2 = (1, 4]
           3 = (4, 7]
           4 = (7, 10]
           5 = (10, 13.66)
           - = Zero, used when the corresponding
               value in a related figure is nonzero
           ▨ = National Grassland areas with
               management choices available

FIGURE 9.3b
Habitat allocations expressed as black-footed ferret carrying capacities under the +20% alternative in year 15.

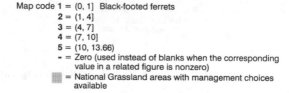

Map code 1 = (0, 1]   Black-footed ferrets
         2 = (1, 4]
         3 = (4, 7]
         4 = (7, 10]
         5 = (10, 13.66)
         - = Zero (used instead of blanks when the corresponding
             value in a related figure is nonzero)
         ▓ = National Grassland areas with management choices
             available

FIGURE 9.3c
Habitat allocations expressed as black-footed ferret carrying capacities under the +20% alternative near equilibrium.

shaded area of the map has been expanded to include the current Grassland colony areas, indicating that choice variables were included for those lands. Although this near-equilibrium analysis appears to extrapolate beyond the supporting data (10–15 years), this may not be the case. If black-footed ferret dispersal is actually a biased diffusion process, the population might be able to take advantage of highly concentrated habitat more quickly than our results would indicate.

The Grassland allocations in figure 9.3c show an interesting pattern resulting from the interaction of two important effects. Many of the sections with the highest potential ferret capacity are left unallocated in order to round out the long, narrow habitat arrangements in portions of the Park. In the model, population losses from fully occupied habitat occur from dispersal across the habitat perimeter into unsuitable areas, and from dispersal into areas already at carrying capacity. For a given amount of habitat, carrying capacity remains constant while dispersal losses into nonhabitat areas can be lowered by reducing perimeter. Allocating circular patterns, which have the smallest perimeter-to-area ratio,

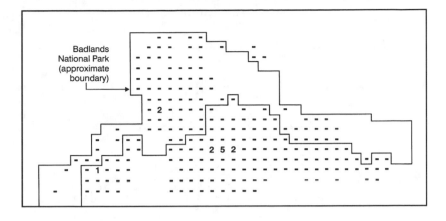

Map code 1 = (0, 1] Black-footed ferrets
2 = (1, 4]
3 = (4, 7]
4 = (7, 10]
5 = (10, 13.66)
■ = Zero (used instead of blanks when the corresponding
value in a related figure is nonzero)

FIGURE 9.4a

The expected number of black-footed ferrets by location under the +20%
alternative released at selected sites.

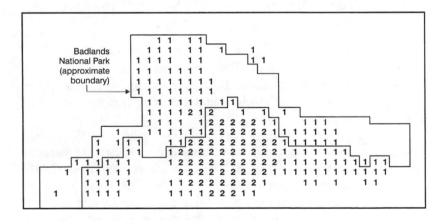

Map code 1 = (0, 1] Black-footed ferrets
2 = (1, 4]
3 = (4, 7]
4 = (7, 10]
5 = (10, 13.66)
■ = Zero (used instead of blanks when the corresponding
value in a related figure is nonzero)

FIGURE 9.4b

The expected number of black-footed ferrets by location under the +20%
alternative in year 7.

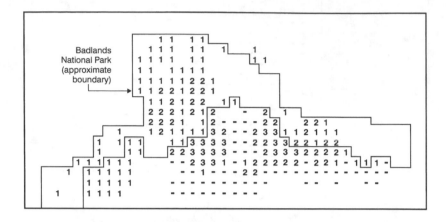

Map code 1 = (0, 1] Black-footed ferrets
          2 = (1, 4]
          3 = (4, 7]
          4 = (7, 10]
          5 = (10, 13.66)
          - = Zero (used instead of blanks when the corresponding
              value in a related figure is nonzero)

FIGURE 9.4c

The expected number of black-footed ferrets by location under the +20%
alternative in year 15.

would minimize losses and maximize retained population given uniformly random dispersal direction. This tendency appears to be compro-
mised somewhat in favor of placing habitat close to as many sections of
the Park as possible.

All the results described so far were obtained using enough different
management variables coupled with constraints on carrying capacity
(rather than land allocations) to provide complete scheduling flexibility.
From a public land use planning point of view, such flexibility may be impractical. Pragmatically, land use planning for a 10- to 15-year period is
often viewed as a process for scheduling a one-time change (if any) in
management, with constraints on land allocations. In some cases, scheduling is not even considered, and all management changes take immediate
effect. Table 9.1 shows the yearly expected ferret populations using these
different approaches to defining habitat management scheduling variables
and constraints for the alternative that allocates 20 percent of the potential
Grassland carrying capacity. Considering the small differences in table
9.1, the use of simpler models (with subsequent greater ease of presenta-
tion and implementation) may not affect results significantly.

TABLE 9.1

The Expected Number of Black-Footed Ferrets in Each Year for the +20 Percent
Alternative Under Three Different Scheduling Formulations

Year	Allocation in Year 1	One-Time Allocation Change in Any Year	Full Scheduling Model
1	9	9	9
2	17	17	18
3	31	31	32
4	51	51	53
5	76	76	79
6	106	107	114
7	139	139	161
8	163	169	177
9	182	183	190
10	196	195	202
11	205	204	212
12	212	210	219
13	217	216	225
14	220	219	229
15	222	222	232
16	224	224	234
17	225	225	236
18	225	226	237
19	226	227	238
20	226	227	238
21	227	228	239
22	227	228	239
23	227	228	239
24	227	228	240
25	227	228	240

Our South Dakota black-footed ferret model is the first application of
this type of dynamic spatial optimization to a real-world problem of
habitat evaluation and management design. With very limited knowledge
of ferret reproduction and dispersal in the wild, the model results must
be regarded as an initial estimate of a lower bound on expected population levels for a given habitat arrangement. The method appears to be
promising in aiding the efficient design of alternative habitat management and reintroduction strategies. The explicit accounting of spatial
patch relationships, rather than relying on measures such as mean inter-
colony distance, is the strong point of the model. Viewing the expected
population estimates from the model as lower bounds provides a useful

contrast to results from habitat complex circumscription methods (such as Biggins et al., 1993), which could probably be viewed as estimates of expected population upper bounds, at least before any qualitative revisions. For species with highly random dispersal behavior, the spatial optimization estimate of expected population should be especially useful.

We must stress that the spatial optimization model is deterministic, and produces estimates of expected population. Stochastic variation, which can be a particularly important consideration at low population levels, is not taken into account, but could be to some degree with the methods discussed in chapter 7. Stochastic variation in net population growth has been accounted for in a model developed by Harris et al. (1989) for estimating black-footed ferret population viabilities as functions of initial population size. Replacing initial population size in their viability model with expected population size from this spatial optimization model would provide a lower bound estimate of expected viability that takes into account stochastic variation in net population growth, an approach that we explore in part IV. For some purposes, this might provide an additional benefit by reducing results to a probability-based index.

Clearly, there is much we do not yet know about black-footed ferret populations in the wild. More information on ferret dispersal could be particularly useful. Consequently, the actual response of the new South Dakota population will probably be somewhat different from model predictions. As more is learned through ferret monitoring over the next several years, we anticipate that the model could be used as part of an adaptive management process (Walters, 1986). The spatial optimization model offers a great deal of flexibility for combining site-specific habitat and population information with management options and constraints.

Chapter 10

A CELLULAR MODEL OF PEST MANAGEMENT

Annual losses in the United States of timber volume to insects and disease are estimated to be 2.4 billion cubic feet (Hamel and Shade, 1987). Thus mitigation of the losses from forest pests is of considerable interest to forest science. We use the term *pest* here to refer primarily to insects, but the chapter could apply to any tree-damaging organism that actively spreads spatially. There are approaches in the optimization literature (such as Reed and Errico, 1987) that account for expected pest-caused losses, as they vary with the timber age class structure created by management choice variables. This allows for a certain degree of optimization in managing pests, but these approaches do not directly account for the population growth and dispersal processes over time and space that enable the pests to spread. This means that predictions of pest-caused losses do not directly take the pest's growth and dispersal behavior into account, and that the optimization model solutions lack spatial detail in terms of where actions should be implemented at each time step. As Polymenopoulos and Long (1990) state, "Any attempt at understanding the stability and persistence of the [pest] populations and communities must take dispersal into account." This chapter explores procedures to optimize the spatial layout of management actions over time, taking these processes into account for a given pest (see Hof et al., 1997). The

This chapter was adapted from J. Hof, M. Bevers, and B. Kent, An optimization approach to area-based forest pest management over time and space, *Forest Science* 43(1) (1997): 121–128, with permission from the publisher, the Society of American Foresters.

modeling approach is described first, followed by a demonstration with a simple example. Similarities between this approach and the one in chapter 8 for wildlife are obvious, but because we are trying to reduce rather than increase pest populations, some very significant differences also surface.

The Model

The approach we present would apply to a pest that is in an active state of growing in population and dispersing spatially (perhaps as a part of a cyclic process). Some pathogens, such as mistletoe (*Arceuthobium* spp.), might behave in this manner over a long time frame as a routine part of their natural history. Our approach typically applies to pests such as pine beetles (*Dendroctonus* spp.), however, which behave this way in an outbreak situation over a much shorter time horizon. Thus the time periods in our approach should be defined differently for different applications.

We again begin by dividing the land into cells, as shown in figure 10.1. As with the time parameters, the cellular layout should be scaled to the pest of concern. As in the last two chapters, we then define choice variables for each cell, where each choice variable represents a scheduled management action. These management actions might involve a va-

21	22	23	24	25
16	17	18	19	20
11	12	13	14	15
6	7	8	9	10
1	2	3	4	5

FIGURE 10.1
A 25-cell planning area.

riety of treatments including poisoning, burning, use of attractants, or sanitation cutting (Amman, 1988). For this discussion we focus our attention on forest treatments rather than direct management actions against the pests themselves.

In order to optimize the spatial layout of management actions while accounting for pest spread over time, some rather specific assumptions regarding pest population growth and dispersal are necessary. Initially, we assume that pest populations in any land cell are quantifiable and determined by the combination of growth and dispersal that spreads the pest from one time period to the next. We assume that the pest has an *r* value that indicates an exponential growth potential per time period, net of pest mortality, in the absence of other limiting factors (Long, 1977). We assume that pests disperse from cell to cell with fixed probabilities that diminish with distance. Skellam (1951) provided a classic treatment that combined a random dispersal pattern (see also Okubo, 1980) with exponential and logistic population growth models. More recently, Liebhold et al. (1995) used this combination and state that "there has generally been good congruence between predictions of this model and observed rates of spread of most exotic organisms." The Skellam model assumed radial dispersal, but we allow factors such as prevailing wind and topography (Paine et al., 1984; Long, 1977) to concentrate the probabilities in certain directions. Thus the population of each cell in any time period after the initial condition is determined (in a manner similar to the approach in chapters 8 and 9) as follows. First, the *r* value is applied to each cell's pest population in the previous time period to determine the unconstrained net growth in population between time periods. Then, it is assumed that the pest has a fixed probability of remaining in the given cell and of dispersing to each other cell between time periods, and that those probabilities decrease with distance according to a known probability density function. This density function is defined such that the sum of all these probabilities in all directions is equal to 1, so that we account for all the pest population.

Thus the expected value of any cell's population that disperses to any other cell between one time period and the next is the appropriate probability times the previous population of the source cell expanded by the *r* value. Conversely, the expected value of the population in each cell in any time period can be calculated by summing, across all cells, the products of the associated probabilities and the previous time period's populations expanded by the *r* value. It is assumed that any pests that leave the area are no longer included in the planning problem, and that no pests enter the planning problem from external sources (although influxes could be accounted for as constants). It is also important to note that we are assuming that the dispersal probabilities are a function only of distance and direction, not of forest characteristics or population levels. We also assume that the management action to be laid out over time is sufficiently effective to reduce the pest population (where it is applied) to preoutbreak (endemic) levels for the remainder of the planning period.

Assuming that we wish to minimize the total pest population over the planning period, a formulation to solve such a problem would be as follows:

Minimize

$$\sum_{t=1}^{T}\sum_{h} S_{ht},\qquad(10.1)$$

subject to

$$\sum_{j=1}^{T} X_{jh} \leq 1 \qquad \forall\, h,\qquad(10.2)$$

$$S_{h0} = N_h \qquad \forall\, h,\qquad(10.3)$$

$$S_{ht} + \sum_{j=1}^{t} a X_{jh} \geq \sum_{n} g_{nh}\left[(1+r)S_{n(t-1)}\right] \qquad \forall\, h\qquad(10.4)$$

$$t = 1,\ldots,T,$$

$$S_{ht} \geq 0 \qquad \forall\, h\qquad(10.5)$$

$$t = 1,\ldots,T,$$

where

h indexes cells, as does n,

t indexes the time periods $(0, 1, \ldots, T)$, as does j $(1, \ldots, T)$,

T = the number of management time periods following $t = 0$ (initial condition),

$X_{jh} \in \{0, 1\}$ = an integer choice variable for the hth cell indicating execution (1) or nonexecution (0) of the management action in time period j,

S_{ht} = the expected population (above endemic levels) of the pest in cell h at time period t,

a = an arbitrarily large constant,

g_{nh} = the probability that the pest will disperse from cell n in any time period to cell h in the subsequent time period. This includes a probability for $n = h$, thus $\sum_h g_{nh} = 1$ for each n,

r = the pest r value, and

N_h = the initial $(t = 0)$ pest population in cell h.

Equation (10.2) forces the selection of no more than one time for execution of the management action for each cell. The first time period ($t = 0$) is used simply to set initial conditions. Equation (10.3) sets the initial population numbers for the pest by cell. Constraint set (10.4) limits each cell's pest population in each time period according to the growth and dispersal from other cells *and itself* in the previous time period. If a management action is selected, then for that time period on, constraint set (10.4) is disabled and the population reverts to an endemic (preoutbreak) level ($S_{ht} = 0$) as the objective function is minimized. The constraint set (10.4) is disabled after any X_{jh} is set equal to 1 because, for each time period t, if any X_{jh} for $j \leq t$ is equal to 1 (not 0) it is multiplied by the arbitrarily large a, meeting the inequality in (10.4) even if S_{ht} is 0. Thus actual pest population growth is determined by a combination of potential growth and dispersal and spatially located limiting management actions. It is important to note that whenever (10.4) is not binding for a cell, the excess pest population assumed to disperse into that cell is lost. Economic objective functions that account for the value of harvested timber (net of pest-caused losses) and the cost of the management treatments could certainly also be specified (as long as they encourage reductions in pest population). This formulation is different from the ones in chapter 8 primarily because we wish to minimize the dispersing population rather than maximize it. This actually creates a much more challenging problem to solve than the one in chapter 8.

In many cases, especially outbreak situations, the ability of the host to support a given pest (which we call host capacity) is eventually exhausted by the pest itself, for one reason or another (see Cole and McGregor, 1983). We model this as a threshold effect that is triggered by the accumulated presence of the pest over time. One way to model this effect, that is similar to the previous model, would be to modify the constraint set as follows:

$$\sum_{j=1}^{T} X_{jh} \leq 1 \qquad \forall\, h, \tag{10.6}$$

$$S_{h0} = N_h \qquad \forall\, h, \tag{10.7}$$

$$S_{ht} + \sum_{j=1}^{t} a X_{jh} + M \cdot F_{ht} \geq \sum_{n} g_{nh}\big[(1+r)S_{n(t-1)}\big] \qquad \forall\, h \tag{10.8}$$

$$t = 1, \dots, T,$$

$$S_{ht} \geq 0 \qquad \forall\, h \tag{10.9}$$

$$t = 1, \dots, T$$

$$F_{ht} \leq 1/\text{TH}_h \sum_{j=1}^{t} S_{hj} \qquad \forall\, h \qquad (10.10)$$

$$t = 1, \ldots, T,$$

$$S_{ht} \leq \sum_{n} g_{nh}\left[(1+r)S_{n(t-1)}\right] \qquad \forall\, h \qquad (10.11)$$

$$t = 1, \ldots, T,$$

where

$F_{ht} \in \{0, 1\}$ = an accessory variable defined such that it is 0 if (10.8)
  is binding and 1 otherwise,

M = an arbitrarily large constant, and

TH_h = the limit of host capacity, over time, for cell h.

At the point where the accumulated presence of the pest over time just equals TH_h in each cell h, constraint (10.8) is rendered nonbinding for that cell because F_{ht} is allowed, by constraint (10.10), to become 1. In the marginal time period, S_{ht} solves at a level just adequate to meet (10.10) with $F_{ht} = 1$, and the remaining S_{ht} in subsequent time periods are allowed to go to 0. Constraint (10.11) ensures that falsely increased S_{ht} levels, which might allow a premature exhaustion of host capacity, are prevented. Note that for a given number of cells and time periods, this model requires twice the number of integer variables as the previous model.

If host capacity is limited, there may be a more powerful way of looking at the problem that does not increase the number of integer variables. If it is possible to detect the trees or areas that the pest has already exhausted (which may require a very intensive monitoring effort), then sanitation management actions might only be effective (or needed) on the remaining trees or areas. This would take advantage of the limited host capacity to minimize the treated area necessary for a given reduction in pest population. This problem could be formulated as follows:

  Minimize

$$\sum_{t=1}^{T} \sum_{h} S_{ht}, \qquad (10.12)$$

subject to

$$S_{h0} = N_h \qquad \forall\, h, \qquad (10.13)$$

$$S_{ht} + M \cdot G_{ht} \geq \sum_{n} g_{nh}\left[(1+r)S_{n(t-1)}\right] \qquad \forall\, h \qquad (10.14)$$

$$t = 1, \ldots, T,$$

$$S_{ht}, Y_{ht} \geq 0 \qquad \forall h \tag{10.15}$$
$$t = 1, \ldots, T,$$

$$G_{ht} \leq 1/\mathrm{TH}_h \sum_{i=0}^{t} S_{hi} + \sum_{j=1}^{t} Y_{hj} \qquad \forall h \tag{10.16}$$
$$t = 1, \ldots, T,$$

$$\sum_{t=0}^{T} S_{ht} + \mathrm{TH}_h \sum_{t=1}^{T} Y_{ht} \leq \mathrm{TH}_h \qquad \forall h, \tag{10.17}$$

$$S_{ht} \leq \sum_{n} g_{nh}\left[(1+r)S_{n(t-1)}\right] \qquad \forall h \tag{10.18}$$
$$t = 1, \ldots, T,$$

where
Y_{ht} = the proportion of cell h that is treated with the sanitation management action in time period t and
$G_{ht} \in \{0, 1\}$ = an accessory variable defined such that it is 0 if (10.14) is binding and 1 otherwise.

All other variables are as previously defined.

With this formulation, constraint (10.14) becomes nonbinding in each cell when the accumulated pest presence *plus* the accumulated area treated just equals TH_h (constraint 10.16). The new management choice variable, Y_{ht}, is continuous and affects pest populations only if the treated area is not already exhausted. The new integer variable, G_{ht}, serves to trigger host exhaustion in constraint (10.14), through constraint (10.16). Constraint set (10.17) limits the pest population to be less than the host capacity minus its reduction from treatment. Constraint set (10.18), like (10.11) in the previous model, prevents a premature exhaustion of host capacity.

An Example

The Problem

For a simple demonstrative example, we assume a problem where a pest outbreak has just begun, and the management objective is to contain it as well as possible. Let us assume that the prevailing winds and topography cause the pest to disperse in a northeast direction, and the initial outbreak is in cell 1 in figure 10.1. The planning area is thus defined according to the expected direction of pest dispersal. We assume an initial pest population of 2 units in cell 1, and that the only feasible management action is sanitation cutting, which reduces the pest population to endemic levels ($S_{ht} = 0$) in all cells. We assume that the duration of the outbreak is expected to be about 4 years. We thus include four 1-year time periods: one that defines the initial conditions and three that are subject to management treatment. A yearly r value of 9 is assumed for the pest such that the unconstrained growth between time periods is a factor of $1 + r = 10$. The g_{nh} coefficients are set at 0.3 for $n = h$, 0.15 for the three cells that are north and east of h, 0.05 for the five cells that are one removed and still north and east from h, and 0.0 for all cells that are farther than that from cell h. Notice that $.3 + (3 \times .15) + (5 \times .05)$ is equal to 1, accounting for all dispersing pests.

For this demonstration, we built two models. The first one (a) is based on constraints (10.2)–(10.5) and has no limit on host capacity. The second one (b) is based on constraints (10.13)–(10.18) and assumes limited host capacity. For this second model, we assume $TH_h = 8$ units for all cells.

Results

We initially solved model (a) with the total sanitation cut in each time period $(\sum_h X_{jh})$ constrained to 0. The resulting pest population for the four time periods was as follows:

t	Population
0	2
1	20
2	200
3	1897.5

It basically grows at the rate of the r value until the pest starts leaving the planning area in time period 3. This reflects the assumption that the host does not limit the pest population. This would suggest that the pest outbreak has no natural termination point, which may not be realistic.

We next solved model (b) with the total cut in each time period $(\sum_h Y_{ht})$ constrained to 0. The resulting pest population for the four time periods was as follows:

t	Population
0	2
1	20
2	94.5
3	83.5

With this model the population is significantly affected by the limited host capacity after time period 1. In fact, pest population declines after time period 2, which is consistent with the notion that the outbreak will only last about 4 years even if no management actions are taken. We also experimented with removing constraint (10.18) from this model. In this case, the same solution was obtained with or without (10.18), but the presence of constraint (10.18) actually reduced solution time by about 50 percent. Thus we retained it in all model (b) solutions discussed hereafter.

Table 10.1 and figures 10.2–10.5 present the optimization results when varying amounts of sanitation cutting are allowed. In figure 10.2, cutting was limited to eight cells per management time period. It should be noted that if nine cells are allowed to be cut in time period 1, the pest population can be reduced to 0 immediately by cutting cells 1, 2, 3, 6, 7, 8, 11, 12, and 13. In model (a) (figure 10.2a), cell 12 is left uncut until time period 2, and then a second block of cells (12, 14, 17, 18, 19, 22, 23, 24) is cut to reduce the pest population to 0 (table 10.1). Having to leave one cell uncut in time period 1 necessitates cutting a block of cells in time period 2 to eliminate the outbreak. In figure 10.2b, the initial pest population of 2 in time period 0 uses up .25 of the limited host capacity, so it is necessary to harvest only .75 of cell 1 in time period 1 to reduce the population in that cell to 0. This allows a partial cut of cell 3 in time period 1, but this actually does not affect the objective function value (table 10.1). The objective function value for model (b) is always less

TABLE 10.1

Numerical Solution Values for Figures 10.2–10.6 ((a) refers to model formulation (10.2)–(10.5), (b) refers to model formulation (10.13)–(10.18))

	Figure 10.2		Figure 10.3		Figure 10.4		Figure 10.5		Figure 10.6
	(a)	(b)	(a)	(b)	(a)	(b)	(a)	(b)	
Objective function value	1.0	1.0	4.0	4.0	11.5	10.25	5.0	5.0	0.0
Pest population $t = 1$	1.0	1.0	2.0	2.0	10.0	6.0	5.0	5.0	6.0
$t = 2$	0.0	0.0	2.0	2.0	1.5	3.5	5.0	4.5	31.5
$t = 3$	0.0	0.0	0.0	0.0	0.0	0.75	5.0	5.0	0.0
Sanitation cut limit by time period	8	8	7	7	6	6	6	6	6

a

-	2	2	2	-
-	2	2	2	-
1	2	1	2	-
1	1	1	-	-
1	1	1	-	-

b

-	-	3:1	-	2:1
3:1	-	3:1	3:1	3:1
1:1	1:1	1:1	2:1	2:1
1:1	1:1	1:1	2:1	2:1
1:.75	1:1	1:.25 2:.625	2:1	2:1

FIGURE 10.2
Optimal sanitation cut layout for minimizing total pest population with no host capacity limit (a) and with a host capacity limit of 8 units (b), with cut limits of eight cells per time period. The numbers in (a) are time of cut and the numbers in (b) are time of cut: portion of cell cut.

than or equal to that for model (a) because it is not necessary to harvest areas that are already host-exhausted. They are equal in the figure 10.2 solutions because there are no cells in time period 1 where the accumulated sum of cutting and pest presence is within .25 of the host capacity (8) and because total pest population is reduced to 0 in time period 2. In time period 2, it is necessary to harvest only .625 of cell 3 because the cut in time period 1 (.25), plus the pest presence in time period 1 (1 ÷ 8) leaves only .625 of the cell left before host capacity is exhausted. The cuts in time period 3 (and the cut in cell 25 in period 2) are actually unnecessary because the pest population is reduced to endemic levels in time period 2 with the cuts in cells 3, 4, 5, 9, 10, 14, and 15. Constraint (10.17) does not prevent this; it only prevents exhausted cells from being cut unnecessarily. We present this solution as obtained to point out this caution, and suggest that in practice the analyst would probably want to constrain the population figures obtained in these types of solutions and then minimize cutting in a rollover solution to avoid unnecessary cuts. It should also be noted that with all the solutions presented, alternative optima are quite likely.

In figure 10.3, cutting in each time period is limited to seven cells. In figure 10.3a, two cells (11 and 12) now have to be left uncut in time period 1 that then disperse pests into time period 2. These cannot be completely eliminated in time period 2 by cutting only seven cells, so pest populations do not return to endemic levels until time period 3 (table 10.1). Cuts are now needed in all time periods, and the total pest population over three time periods is now 4 units (table 10.1). In figure 10.3b, it is still not possible to capitalize on the additional flexibility of model (b), so the objective function value is also 4. The solutions in figures 10.3a and 10.3b are again quite similar, but are mirror images (again note the possibility of alternative optima).

In figure 10.4, sanitation cutting in each time period is limited to six cells. In figure 10.4a, the strategy is to contain the pest to the row of cells 1, 6, and 11 in time period 1, which then contains it in the northeast corner of the planning area in time period 2. It is particularly interesting to note that it is optimal not to cut cell 1 in time period 1 so as to pursue this strategy. The solution in figure 10.4b is similar (though a mirror image), except that it is optimal to cut in cell 1 in time period 1. We obtained an alternative optimum, in fact, that cut .75 of cell 1 in time period 1 (with compensating adjustments elsewhere), thus eliminating the pest in cell 1 in time period 1. In figure 10.4, the extra flexibility in model (b) allows an objective function value that is 1.25 units less than that for model (a) (table 10.1). The added flexibility in model (b) allows

a

3	2	2	3	3
2	2	2	3	3
2	2	1	3	3
1	1	1	-	-
1	1	1	-	-

b

-	3:.75	3:1	3:1	3:1
-	-	3:.9375	2:.5 3:.4375	3:.9375
1:1	1:1	1:1	2:1	2:1
1:1	1:1	2:.875	1:.25 2:.75	2:1
1:.75	1:1	2:.875	2:1	3:.9375

FIGURE 10.3

Optimal sanitation cut layout for minimizing total pest population with no host capacity limit (a) and with a host capacity limit of 8 units (b), with cut limits of seven cells per time period. The numbers in (a) are time of cut and the numbers in (b) are time of cut: portion of cell cut.

a				
3	3	3	3	3
2	2	2	3	-
2	1	1	-	-
2	1	1	-	-
2	1	1	-	-

b				
-	3:.4375	-	-	3:1
-	-	2:1	3:.9375	3:.9375
1:1	1:1	1:1	2:1	3:.875
1:1	1:1	2:.875	2:1	3:.875
1:.625	1:.375 2:.25	2:.875	2:1	3:.9375

FIGURE 10.4

Optimal sanitation cut layout for minimizing total pest population with no host capacity limit (a) and with a host capacity limit of 8 units (b), with cut limits of six cells per time period. The numbers in (a) are time of cut and the numbers in (b) are time of cut: portion of cell cut.

the solution to pursue the same basic strategy as in figure 10.4a, but to also attack cell 1 in time period 1. The resulting population schedule in table 10.1 has four fewer units in time period 1 for model (b) than for model (a). With even more restrictive constraints on harvesting, this distinction between models (a) and (b) might increase further. However, we encountered debilitating increases in solution time with these more restrictive constraints. It should be noted before proceeding that the cut of .4375 of cell 22 in time period 3 in figure 10.4b is, again, an example of unnecessary cutting indicated in solution.

In viewing the numeric results for figure 10.4 in table 10.1, the schedule of pest populations indicates large initial populations followed by steady reductions. These results reflect the objective function of minimizing the total population across time periods. In some cases, the highest periodic pest population may be the main concern, rather than the total population over time. To address this situation, it may be desirable to minimize the maximum time period's pest population (MINMAX) by replacing the previous objective function with the following:

Minimize

$$\lambda, \tag{10.19}$$

subject to

$$\lambda \geq \sum_h S_{ht} \quad \forall\, t \tag{10.20}$$

in either the (a) or (b) model (all previously defined constraints are retained).

Figure 10.5 presents solutions for models (a) and (b), with sanitation cutting in each time period limited again to six cells, but with (10.19) and (10.20) replacing the previous objective function. From table 10.1, the model (a) solution now has half the pest population in time period 1 (compared to figure 10.4a results), which is also the pest population in time periods 2 and 3. This solution thus increases total pest population over all three time periods by 3.5 units, in order to decrease the maximum time period's population by 5 units. The decrease in time period 1 population is accomplished largely by applying the sanitation cut to cell 1 in time period 1. The spatial applications of management in figures 10.4a and 10.4b seem quite similar to those in 10.5a and 10.5b, but the subtle differences yield a very different schedule of (and total of) pest populations. These differences between figures 10.4b and 10.5b (see table 10.1) are less pronounced than those between 10.4a and 10.5a. In comparing the results for figures 10.4b and 10.5b, the reduction in time period 1 population of 1 unit was accomplished by allowing total popua

-	-	-	-	-
-	-	3	3	3
1	1	1	2	3
1	1	2	2	3
1	1	2	2	3

b

3:.9375	3:.84375	3:.875	3:.9375	3:.6875
2:1	2:1	2:.9375	2:.9375	3:1
2:.875	2:.875	1:1	-	3:.71875
1:.25 2:.375	1:1	1:1	-	-
1:.75	1:1	1:1	-	-

FIGURE 10.5

Optimal sanitation cut layout for minimizing the maximum periodic pest population with no host capacity limit (a) and with a host capacity limit of 8 units (b), with cut limits of six cells per time period. The numbers in (a) are time of cut and the numbers in (b) are time of cut: portion of cell cut.

lation over all three time periods to increase by 4.25 units. In the results for Figures 5a and 5b, it should be pointed out that nonbinding population levels may not be at their minimums. Additional rollover solutions may be necessary if it is desired to take the slack out of all time periods' populations.

In figures 10.4b and 10.5b, some pest population is still present in time period 3. In some cases, minimizing the pest population at some target time period (such as the last one in the model) may be of paramount importance. Figure 10.6 presents a solution to model (b), again with cuts limited to six cells per time period, but minimizing only the third time period's pest population. With this objective function, the strategy is quite different: Contain the pest to a path (cells 1, 7, 13, 19, and 25) straight through the planning area and into the north part of the planning area in the first period, and then either eliminate the pest or let it move out of the planning area in subsequent time periods. This might suggest the importance of defining the planning area relative to starting conditions, the pest's ability to disperse, and the time frame involved.

1:.4375 2:.5	2:.875	3:.8125	3:.875	3:.9375
1:.8125	3:.4375	2:.375	3:.5625	3:.9375
3:.5	-	2:.875	2:1	2:1
1:1	2:.375	1:1	2:1	-
1:.75	1:1	1:1	-	-

FIGURE 10.6
Optimal sanitation cut layout (time of cut: portion of cell cut) for minimizing third time period pest population with a host capacity limit of 8 units and cut limits of six cells per time period.

Any number of other objective functions might be used, including economic ones and those that account for a nonlinear relationship between pest populations and the damage they cause. The focus of this example has been capturing the spatial and dynamic spread of pests in an optimization framework.

Discussion

The example presented here is clearly not a real-world problem and no generalizations are intended. Its purpose is limited to demonstrating the model formulations in a simple example that is easily interpretable. Even with this simple example, obtaining solutions was difficult for the results reported (with typical solution times on a 486/66 microcomputer of more than 24 hours) and impossible for more restrictive cutting constraints (2 weeks being inadequate to find a global optimum). Also, the nature of the formulations causes the bounds on optimality (that come from the linear program solutions in the branch-and-bound algorithm) to be so far from actual optimality that they are not useful. As the hardware and integer programming solvers continue to improve, more problems will be solvable. Also, we have begun to work on formulations for the problem where management actions treat (exterminate) the pest directly, as opposed to treating the host. With no limits to host capacity included, this problem appears to be analyzable, but this approach needs further development. In the meantime, this chapter has presented some exploratory ideas and demonstrated that at least some problems might be analyzable in the fashion discussed. Because this model must necessarily predict a process that is largely random (with fixed probabilities), it would best be applied in an adaptive management setting, where outbreak behavior and responses to management actions are monitored, with results fed back into additional analyses and adjustments in management strategy.

Chapter 11

A NESTED-SCHEDULE MODEL OF STORMFLOW

Public and scientific concerns over the cumulative effects of multiple forest disturbances within a watershed appear to be growing. It is clear that forest treatments, such as timber harvesting or prescribed burning, can significantly increase onsite storm runoff (see Robichaud and Waldrop, 1994) as well as annual water yield (see Troendle and King, 1987), particularly on steep slopes. In a recent report commissioned by the president of the United States regarding management options for forest-dependent species and fisheries at risk in the Pacific Northwest (FEMAT, 1993), federal land management agencies strongly recommended the use of watersheds as a basic management planning unit due to cumulative effects of forest projects. Some of the direct physical effects, such as channelization or sediment transport and deposition, can often be attributed to changes in land use and the resulting peak flows and total volumes of short-term runoff from storm events (see Janda et al., 1975). In some cases the relationship with stormflow may be less clear. For example, cumulative effects of watershed changes on stream water temperatures have been attributed to increases in stormflow (Pluhowski, 1970), but are also affected by loss of streamside shading (Brown and Krygier, 1970).

Stormflows within and downstream from a drainage system result from a complex spatial and temporal combination of runoffs strongly af-

This chapter was adapted from M. Bevers, J. Hof, and C. Troendle, Spatially optimizing forest management schedules to meet stormflow constraints, *Water Resources Bulletin* 32(5) (1996): 1007–1015, with permission from the publisher, the American Water Resources Association.

fected by forest cover conditions. Coincidences of peak flows from a network of drainages would be of particular importance when they occur. Estimating local stormflows and routing combined runoff is a challenging and rapidly developing field of endeavor (see Julien et al., 1995). For example, runoff attenuation over distance as a function of detention and channel storage (Horton, 1935) must be taken into account for each tributary in order to estimate the timing of confluent stormflows.

An important watershed management challenge with additional difficulty that must be spatially addressed is the longer-term prescriptive problem of scheduling and arranging forest treatments (which determine forest cover conditions over time) in a watershed to strategically meet both forest management and stormflow management objectives. This chapter presents an exploratory method for addressing this spatial scheduling problem (see Bevers et al., 1996).

The Spatial Optimization Approach

Identifying the location and timing of forest treatments to best meet forest and stormflow management objectives suggests that a spatial forest scheduling model be combined with a stormflow simulation and routing model. Combining two such models is not straightforward because time scales appropriate for modeling forest growth and management are substantially longer than the time scales appropriate for routing stormflows. Forest management is often modeled over a series of discrete planning periods that are years or decades in length, whereas stormflows are typically simulated over a series of discrete time steps that may be only minutes or hours in length. Another difficulty is that over a 50- or 100-year forest management planning horizon, timing of storm events cannot be predicted. We attempt to resolve these challenges with a nested time schedule. Our approach is to select one or more storm events of interest for watershed planning purposes and model the stormflows from each of those storms over a sequence of appropriately short time steps within each forest management planning period. By setting upper bounds on either the water storage or outflow amounts per storm event time step, forest treatments can be scheduled over a long time horizon across the watershed so as to achieve forest management objectives while meeting the stormflow management objectives regardless of when the storm events occur. Any problems with spatially synchronized runoff from storms can then be managed with the long-term management actions that determine patterns of forest cover.

As with so many other problems analyzed in this book, it is necessary to define our decision choices about the spatial arrangement and timing of forest management activities with mixed-integer programs using binary variables. As we have seen, these mixed integer programs can be very difficult to solve, even when restricted to linear systems of equations. Consequently, for this exploratory effort it is useful to simplify the hydrologic portion of the model into linear equations, taking care to specifically retain stormflow sensitivity to changes in forest cover.

We assume an overall forest management objective to maximize total discounted net returns from timber harvests, subject to constraints on peak stormflows. Using a watershed area subdivided into forest management units (microcatchments) whose boundaries define discrete stream channel sections (reaches), the spatial optimization model is as follows:

Maximize

$$\sum_h \sum_{k \in K_h} v_{hk} X_{hk}, \tag{11.1}$$

subject to constraints on forest management units

$$\sum_{k \in K_h} X_{hk} = 1 \qquad \forall\, h \qquad X_{hk} \in \{0, 1\}, \tag{11.2}$$

$$W_{nhp1} = \sum_{k \in K_h} a_{nhkp} X_{hk} \qquad \forall\, n,\, h,\, p, \tag{11.3}$$

$$
W_{nhp(t+1)} = (1 - b_{nh}) W_{nhpt} - O_{nhpt} + \sum_{i \in I_h} g_{ih} O_{nipt}
$$
$$
+ \sum_{k \in K_h} u_{nhkpt} X_{hk} \qquad \forall\, n,\, h,\, p \qquad t = 1, \ldots, T_n - 1, \tag{11.4}
$$

$$O_{nhpt} \le r_{nhps} W_{nhpt} + m\left(1 - \sum_{k \in K_{hs}} X_{hk}\right) \quad \text{and}$$

$$O_{nhpt} \ge r_{nhps} W_{nhpt} - m\left(1 - \sum_{k \in K_{hs}} X_{hk}\right) \tag{11.5}$$

$$\forall\, n,\, h,\, p,\, t,\, s \qquad K_{hs} \subset K_h$$

and constraints on stream channel sections

$$W_{njp1} = q_{nh}, \tag{11.6}$$

$$W_{njp(t+1)} = (1 - c_{nj} - b_{nj})W_{njpt} + \sum_{d \in D_j} c_{nd} W_{ndpt}$$

$$+ \sum_{i \in I_j} g_{ij} O_{nipt} + z_{njt} \qquad \forall\, n,\, j,\, p,\, t, \qquad (11.7)$$

where

h indexes forest management units,

j indexes stream channel sections,

k indexes members of the set K_h of forest treatment schedules available for each management unit h,

p indexes long-term planning periods (years, decades) over some planning horizon from 1 to P for modeling changes in forest cover resulting from scheduled treatments and natural growth,

t indexes short-term time steps (minutes, hours) over some storm event from 1 to T_n for modeling stormflows within each planning period,

n indexes storm events selected for a given model,

i indexes the set I_h or the set I_j of adjacent, upslope management units whose outflows contribute directly to the inflows for either management unit h or stream channel section j, respectively,

d indexes the set D_j of upstream channel and tributary sections whose discharge becomes direct inflow to stream channel section j,

X_{hk} = the binary choice variable (assigned 0 or 1) for each forest management unit and treatment schedule combination,

v_{hk} = the expected present net return from harvesting forest unit h according to schedule k,

W = a state variable representing the amount of water (storage) contributing to stormflow in management unit h or in stream channel section j at time step t of planning period p given storm event n,

a = an initial water amount assumed for management unit h in planning period p if treatment schedule k has been selected, given storm event n,

b = the proportion (often negligible during a storm event) of water lost per time step through percolation, evaporation, or transpiration from management unit h or stream channel section j,

O = a variable representing the outflow from a management unit (h or i) during time step t of planning period p given storm event n,

g_{ih} = the proportion of outflow from management unit i expected as direct inflow to management unit h,

u = the net amount of exogenous water from precipitation, snowpack, and other outside sources (after accounting for interception, snowmelt, and similar processes) expected to enter management unit h during time step t of planning period p given storm event n and forest cover conditions resulting from the selection of treatment schedule k,

r = the proportion of water expected to flow out from management unit h to other management units or channel sections during each time step of planning period p given selection from the treatment schedule subset K_{hs} (of set K_h), whose members are expected to produce similar forest conditions in that period,

q = an initial water state base amount assumed for stream channel section j given storm event n,

m = a constant large enough to provide slack for inactive outflow constraints,

c = the proportion of water in a stream channel section (j or d) expected to flow downstream into the next channel section during each time step, and

z = the amount of exogenous water from direct precipitation and baseflow sources expected to enter stream channel section j during time step t given storm event n.

The objective function for maximizing present net revenue in equation (11.1) schedules positively valued timber harvests as early in the planning horizon as possible (subject to constraints) to reduce discounting of expected returns. Forest treatments are then strategically deferred by setting upper bounds on either the water state variables (W) or the outflow variables (O) to control stormflow cumulative effects.

Equation (11.2) requires that one and only one forest treatment schedule be selected for each management unit. Equations (11.3) and (11.6) set the initial water conditions, by storm event, for each forest management unit and stream channel section, respectively. Because changes in forest cover can affect antecedent moisture levels, equation (11.3) makes the initial water amount a function of the planning period and the treatment schedule selected. No attempt is made here to simulate antecedent conditions involving water deficits, such as soil moistures below field capacity. Such antecedent conditions would not ordinarily be of interest in problems aimed at limiting peak stormflows.

Equation (11.4) provides a water balance equation for each time step of every storm, by planning period, for all forest management units in the watershed. The schedule of water added to each management unit from net precipitation, snowmelt, and other sources is specifically modeled as a function of the storm event time step, as well as the planning period and selected forest treatment schedule. Thus this approach could be used to model rain-on-snowpack storm events if it is assumed that on-site snowmelt rates are unaffected by the amount of inflow from other management units and if it is assumed that variations in snowpack are unaffected by juxtaposition of harvests. Equation (11.7) provides similar water balance equations for all stream channel sections in the drainage system.

Stormflow sensitivity to changes in forest cover is achieved through the paired constraints of equation (11.5). These constraints set the outflow rates from each forest management unit as a function of the treatment schedule selected for that unit. For example, consider the situation in the first planning period for a forest management unit (say, $h = 1$) uniformly covered by old growth, for which a number of clearcut scheduling options (comprising the set K_1) are included. In such a case, only two cover conditions are possible in the first period of the model: old growth or cutover. One subset (K_{11}) of the scheduling options would typically entail harvesting the entire unit (keeping in mind that the X_{hk} are 0–1 variables) in the first period, and the only other possible subset (K_{12}) would include all the scheduling options that postpone harvesting until a later period. If a schedule is chosen from subset K_{11}, then $\sum_{k \in K_{11}} X_{1k} = 1$ so that the large constant m is multiplied by 0. The paired constraints in equation (11.5) then require an equality condition that determines the outflow based on a rate r_{n111} (for storm event n in period 1) that is appropriate for cutover conditions in that management unit. Note, however, that m must be positive and set large enough that the outflow is not influenced by the alternative rate r_{n112} that is intended for old growth cover conditions in the other pair of similar constraints. Naturally, more than two cover conditions would become possible in later planning periods.

Stormwater comes from a number of sources, including surface runoff (Horton, 1935), subsurface flows (Hursh and Brater, 1944), and return flows (Musgrave and Holtan, 1964), all of which appear as highly variable active source areas during a storm event (Hewlett and Hibbert, 1967; Troendle, 1985). These stormwater sources contribute to dynamic hillslope outflows into streams (see Roessel, 1950). Consolidating stormwater from all sources on each forest management unit into a single state variable with a proportional rate of outflow (which varies only by changes in forest cover) must be viewed as a first-order approximation of a much more complex process. Nonetheless, a time series of proportional outflows approximates a negative exponential process and does a reasonable job of reproducing typical storm hydrographs with inflow, storage, outflow, and summation characteristics similar to those described by Dunne and Leopold (1978).

Two Examples

Two hypothetical cases were tested using this spatial optimization model. For the first case (case A, figure 11.1a), the watershed comprises 24 square 16-ha management units (labeled 01–24), and is bisected by a stream 6 m wide. Four management units adjoin the stream on each side, defining four channel sections (labeled C1–C4). The land is uniformly sloped such that 95 percent of stormflows move directly toward the stream and 5 percent move into adjacent land units parallel to the stream.

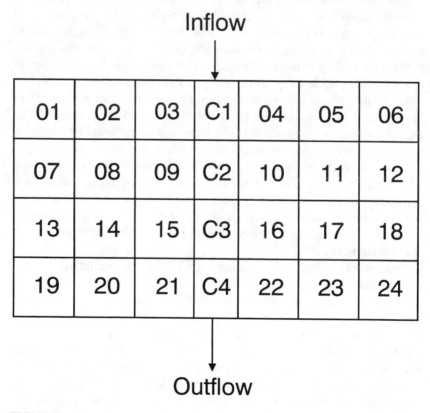

FIGURE 11.1a
Hypothetical old growth forest management units and stream channel sections (not to scale) for watershed case A.

Consequently, small amounts of water leave the watershed from land units 19–24, with the principal discharge emanating from stream channel section C4. Water enters the area only as precipitation (net of interception), and as base streamflow into channel section C1. The land is initially covered uniformly with old growth forest.

Conditions are similar for the second case (case B, figure 11.1b), except that tributaries of negligible width and without baseflow are added to the drainage system. Upper management units (01, 02, 05–08, 11–14, 17–20, 23, and 24) are sloped so that 95 percent of stormflows move di-

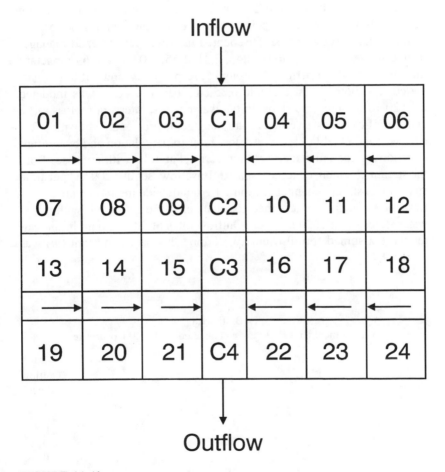

FIGURE 11.1b
Hypothetical old growth forest management units and stream channel sections (not to scale) for watershed case B.

rectly into the tributaries and 5 percent move into adjacent, downslope land units parallel to the tributaries and toward the principal stream. For the 8 management units adjacent to the main stream (03, 04, 09, 10, 15, 16, 21, and 22), 50 percent of stormflows move into the tributaries and 50 percent move directly into the main stream channel sections. The tributaries empty into the upper reaches of stream channel sections C2 and C4. With all land units draining directly into channels, case B provides a much faster and more complex stormflow runoff system. Note that in this case, water can leave the area only as discharge from stream channel section C4.

Treatment schedules for each management unit were designed in both cases to allow a single harvest in any of five 10-year planning periods, as well as a no-harvest choice. Discounted net revenues for each management unit were assumed to be 3.45, 2.31, 1.55, 1.03, and 0.69 monetary units for harvests in periods 1 through 5, respectively, and 0 if left unharvested. For both cases, a single 6.35-cm, 1-hr rainstorm was modeled over 20 10-minute time steps. Initial water states, interception loss rates, and stormflow rates assumed for all management units are shown by forest cover age in table 11.1. Water loss rates to percolation and evapotranspiration for each management unit were set at .001, with no losses from the stream channel sections assumed. Baseflow, set at 1200 m³ per time step, enters stream channel section C1 from outside the model area. Outflow rates of .75 were assumed for channel sections C1–C4, with initial water states set at 1600 m³ each. Outflow rates of .85, with no initial water, were assumed for all tributary sections. The large constant (m) was set at 99,999.

TABLE 11.1
Management Unit Hydrologic Parameters by Forest Cover Age

Age in Decades	Initial Water State (m³)	Interception Loss Rates		Outflow Rates (r)
		$t = 1$	$t = 2, \ldots, 6$	
0	20,000	.10	.02	.009
1	19,200	.30	.10	.006
2	18,400	.40	.14	.005
3	17,600	.50	.18	.004
4	16,800	.60	.22	.003
Old growth	16,000	.80	.36	.001

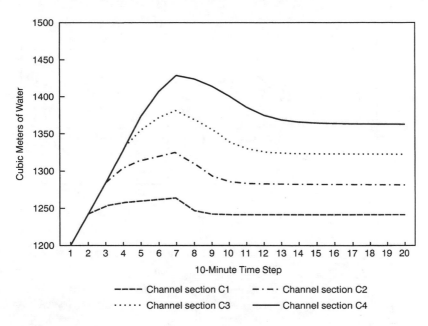

FIGURE 11.2a
Modeled hydrologic response by channel section to a 1-hr, 6.35-cm rainstorm for watershed case A.

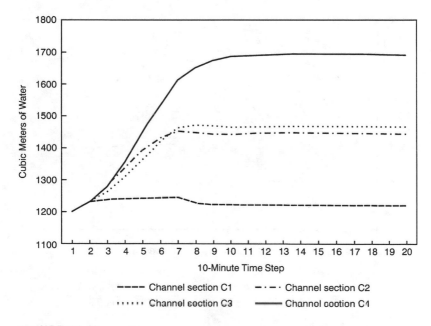

FIGURE 11.2b
Modeled hydrologic response by channel section to a 1-hr, 6.35-cm rainstorm for watershed case B with old growth cover.

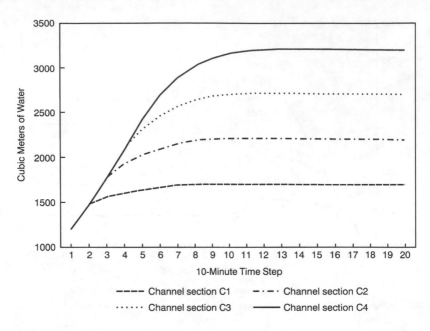

FIGURE 11.3a
Modeled hydrologic response by channel section to a 1-hr, 6.35-cm rain-storm for watershed case A.

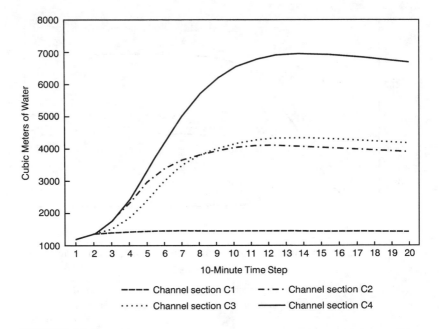

FIGURE 11.3b
Modeled hydrologic response by channel section to a 1-hr, 6.35-cm rain-storm for watershed case B following old growth removal.

Results

Figures 11.2a and 11.2b show hydrographs by stream channel section under the initial old growth cover conditions for cases A and B, respectively. Although both cases begin with baseflow conditions of 1200 m^3 per time step, the peak stormflow increment at the watershed outlet (C4) is more than twice as much (nearly 500 m^3 per time step) for case B as for case A (less than 250 m^3 per time step). Case A peaks much earlier (at $t = 7$) than case B (at $t = 15$), which would result in a longer, flatter tail on the recession limb of the hydrograph for case A if enough time steps were shown. Tributary flows into channel section C2 are responsible for the unusual hydrographs of channel sections C2 and C3 in figure 11.2b. The faster response of the tributaries causes the early portion of the rising limb of the C2 hydrograph to exceed that of the C3 hydrograph. Also, after both hydrographs peak and subside slightly, they exhibit a smaller secondary rise before receding.

Figures 11.3a and 11.3b show the opposite extreme of stormflow response with hydrographs by stream channel section for completely cutover forest conditions. For case A, harvesting all 24 management units in a single time period results (in that period) in peak stormflows (figure 11.3a) more than twice those under old growth cover conditions (figure 11.2a). For case B, cutover peak stormflows (figure 11.3b) are quadrupled over those shown for old growth cover in figure 11.2b. Recovery of the cutover watershed over time is shown by the C4 hydrographs in figures 11.4a (case A) and 11.4b (case B).

Some response characteristics of the simpler watershed for case A are noteworthy. Without an explicit drainage system to rapidly move water from upper and midslope management units, the effects of those units on peak flows in the stream are very small. The hydrographs that result from harvesting all 16 of the upper and midslope management units in a single period are nearly identical to the hydrographs for complete old growth cover conditions in figure 11.2a, as long as the streamside units remain under old growth cover. Conversely, the hydrographs that result from harvesting all 8 streamside management units (leaving old growth cover on the upper and midslope units) are nearly identical to the hydrographs for completely cutover forest conditions (figure 11.3a). In case A, water from the upper and midslope management units must move through additional land units before reaching the stream. The slow outflow rates (and high detention and storage capacity) of those intervening land units in the model appear to buffer the stream from the effects of upslope harvest activities. A more extensive drainage system precludes such results for case B.

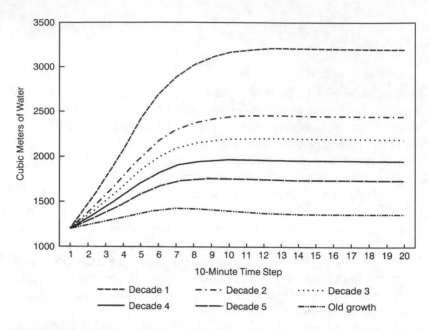

FIGURE 11.4a
Watershed recovery over time as shown by outlet flows from 1-hr, 6.35-cm
rainstorms following old growth removal in period 1 for watershed case A.

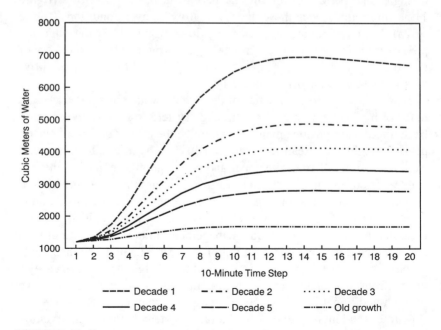

FIGURE 11.4b
Watershed recovery over time as shown by outlet flows from 1-hr, 6.35-cm
rainstorms following old growth removal in period 1 for watershed case B.

As stated earlier, the main purpose for our model is to help decision makers to spatially schedule forest treatments in a manner that simultaneously meets both forest management and stormflow management objectives. We tested this use of the model by maximizing discounted net revenue subject to peak stormflow constraints at the watershed outlet (C4 outflows) of 2250 m^3 per time step for case A and 3750 m^3 per time step for case B.

The resulting harvest schedule for each management unit (coded NH for no-harvest units) is shown in figure 11.5a for case A. The model is able to schedule harvests in all 16 of the upper and midslope management units in the first period, with an overall objective function value of 73.44 discounted monetary units. The problem of meeting the peak flow constraints for that case reduces to a matter of appropriately scheduling the 8 streamside units. Because of the highly uniform nature of the modeled watershed, case A is symmetric about the main stream channel. For that reason, at least one alternative optimum can always be found by simply reversing the results between the left and right halves of the watershed. With the results in figure 11.5a, this would mean reassigning harvest periods so that management unit 09 is harvested in the third decade, units 10 and 15 in the second decade, unit 16 in the fourth decade, unit 21 in the fifth decade, and unit 22 in the first decade. Both of these solutions result in C4 peak flows of about 2068 m^3 per 10 minutes (at $t = 16$) in decade 1, 2234 m^3 per 10 minutes (at $t = 12$) in decade 2, 2182 m^3 per 10 minutes (at $t = 12$) in decade 3, 2164 m^3 per 10 minutes (at $t = 11$) in decade 4, and 2130 m^3 per 10 minutes (at $t = 11$) in decade 5. Peak flow constraints were not precisely binding because of the use of integer management variables. Besides the alternative optima resulting from symmetry, we found several other (nonsymmetric) optimal solutions for case A (each, in turn, with its own symmetric alternative). These solutions result in somewhat different but equally acceptable runoff patterns and peak flows from the solution given in figure 11.5a.

With case B (figure 11.5b), the problem is at a level of watershed complexity for which it would be very difficult to deduce an optimal solution without mathematical programming. An overall objective function value of 48.55 discounted monetary units is achieved with C4 peak flows of 3636 m^3 per 10 minutes (at $t = 15$) in decade 1, 3738 m^3 per 10 minutes (at $t = 15$) in decade 2, 3565 m^3 per 10 minutes (at $t = 15$) in decade 3, 3698 m^3 per 10 minutes (at $t = 14$) in decade 4, and 3680 m^3 per 10 minutes (at $t = 14$) in decade 5. Case B has numerous alternative optima with identical stormflow patterns because watershed conditions are symmetric about each tributary, as well as about the main stream. We did not find any other (nonsymmetric) alternative optima, although one

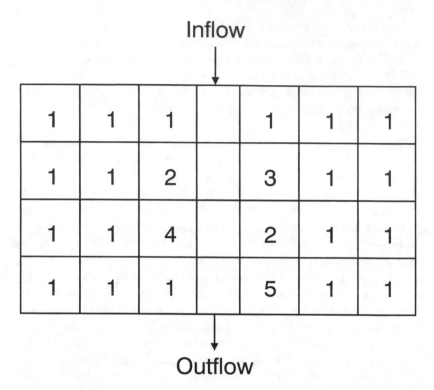

FIGURE 11.5a
Optimal periods of forest treatment by management unit given outlet flow upper bounds of 2250 m³ per 10 minutes for watershed case A.

or more may exist. The results in figure 11.5b show a tendency to defer or eliminate harvests immediately adjacent to the main stream channels, somewhat like the results for case A. Also, the streamside units still appear to have the greatest influence on stream channel stormflows. Units 17, 19, and 23 are notable exceptions, however, and the other upper land units are scheduled for harvest in a mix of planning periods 1 and 2, rather than strictly in period 1. Suitable locations for such scheduling exceptions would be difficult to predict without this sort of spatial optimization model.

Discussion

The spatial optimization approach presented here shows promise for addressing cumulative effects management of forested watersheds. Even

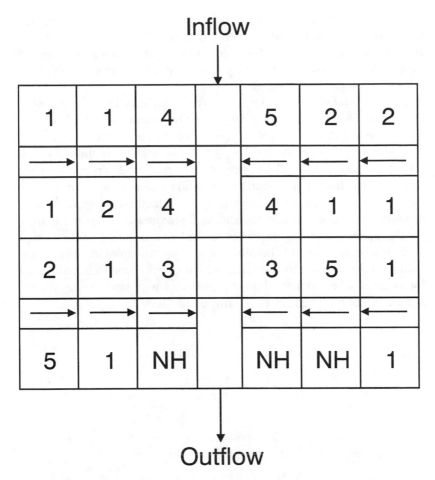

FIGURE 11.5b
Optimal periods of forest treatment by management unit given outlet flow
upper bounds of 3750 m³ per 10 minutes for watershed case B.

with minimal complexity in the hydrologic equations, however, these
models pose serious solution difficulties. The paired constraints of equation (11.5) for setting management unit outflow rates as a function of
forest cover act as switches that reduce the efficiency of branch-and-
bound mixed-integer solution algorithms. Unfortunately, many pairs of
these constraints are required for even a small model in the present formulation. For the results presented, solution difficulties were resolved by
using increasingly refined advanced starts. With a more realistic water-

shed this may prove impractical, and heuristic approaches to estimating solutions for combinatorial problems (Reeves, 1993) may be required. Heuristic techniques would almost certainly be required if more complex hydrologic equations were used, given current mathematical programming and computer technology.

Despite solution difficulties, the two examples clearly demonstrate the potential for spatially scheduling forest treatments to control cumulative watershed effects, as well as the use of mathematical programming for addressing these problems. Using fairly simple but spatially explicit hydrologic modeling resulted in reasonably realistic hydrographs while still allowing formal (nonheuristic) solution procedures to be applied. The use of nested time schedules provides a method for linking forest and stormflow management models, and provides a useful mechanism for designing watershed decisions around one or more storm events. With these constructs, it is possible (at least in principle) to optimize the spatial layout of forest management activities over time so as to create a forest cover schedule that minimizes problems with stormflow given the uncertainty of the timing of the storms themselves.

PART IV

DIVERSITY AND SUSTAINABILITY

Up to this point in the book, our principal focus has been on devising formulations that capture spatial relationships in optimization models. For the most part, we have used fairly simple objective functions to explore the workings of the resultant formulations. To realistically address the use of spatial optimization models for managing ecosystems, however, we must also recognize the importance of the objective functions themselves. In this part, we expand our use of objective functions and extend the discussion of dynamic equilibria that we began in part III.

Forest management optimization models have typically been formulated so as to maximize an economic objective function through allocation of land to various management prescriptions. Recently, considerable interest has been generated in managing ecosystems for more biological objectives. One objective of particular interest is the concept of biodiversity. To allow formulation of biodiversity objectives in management optimization models, the concept must be rigorously defined. Some ecological literature emphasizes different criteria in defining biological diversity for different ecological scales (see Allen and Hoekstra, 1992). Genetic diversity is probably most important for populations where a single species, or a group of species that are fairly similar, is the focus. For communities or small-scale ecosystems, it is diversity between highly differentiated populations or ecosystem components (such as producers, consumers, and decomposers) that is important, and genetic diversity may not be very relevant. For ecological systems at landscape scales, diversity between ecosystems is important and genetic diversity is almost certainly washed out. For biomes, diversity in landscapes is the concern. A bit more simply, Westman (1990) defines biodiversity as species diversity, habitat diversity, and genetic diversity, where species diversity includes measures of both richness and equity (or evenness) of relative abundance in a given area. Chapters 12 and 13 provide a set of approaches that address species diversity and habitat diversity, but genetic diversity is much more elusive.

Several recent papers, most notably Weitzman (1992, 1993) and Solow et al. (1993), have discussed methods for using pairwise measurements of DNA differences or "distances" between species to construct an overall measure of genetic diversity. These methods are motivated by the commonality of these pairwise DNA distances in the genetics literature. It appears that if it is desired to account for any given species's genetic uniqueness, then these pairwise distances are the likely form of usable genetic information that is available. In natural resource optimization, a biodiversity measure would be included as either a constraint or an objective function in a mathematical program, so a closed-form function is needed. Weitzman provides a rather comprehensive theoretical discussion on this topic, and proposes a recursive dynamic programming equation that defines what Solow et al. call a pure diversity measure. It is pure in the sense that it evaluates a given set of species (with probabilities of survival) relative only to itself. Solow et al. note that "a practical algorithm for its evaluation is not currently available unless the number of elements of X (number of species) is small." It is also clear that the Weitzman approach requires recursion, implying that it cannot be captured with a closed-form function. Solow et al. suggest a preservation measure that evaluates a given set of species (with probabilities of survival) in terms of the loss in genetic distance relative to the current set of species. The Solow et al. approach does not require recursion, but does involve a large, combinatorial calculation that also makes its inclusion as a closed-form function in an optimization model intractable. Given the nature of the genetic information available, no practical method for constructing tenable genetic diversity objective functions currently exists.

Even if a method were available for using the pairwise genetic difference data to construct an overall objective function, it would still probably suffer from the limitations of pairwise comparisons. For example, with three species, A, B, and C, there are seven possible events related to uniqueness of a given trait selected at random. Either the trait is unique to one of the three species, or it is unique to one of three possible pairs of species, or it is shared by all three species. If N_A, N_B, and N_C are the total number of genetic traits found in species A, B, and C, respectively, $N_{A \cup B}$ is the total number of unique traits represented by species A and B collectively, $N_{A \cap B}$ is the number of unique traits shared by species A and B, and so on, then the probability that a trait is unique to a single species (A, for example) is

$$\frac{N_A - N_{A \cap B} - N_{A \cap B} + N_{A \cap B \cap C}}{N_{A \cup B \cup C}}, \qquad (\text{IV.1})$$

or

$$\frac{N_A - N_{A \cap B} - N_{A \cap C} + N_{A \cap B \cap C}}{N_A + N_B + N_C - N_{A \cap B} - N_{A \cap C} - N_{B \cap C} + N_{A \cap B \cap C}}. \qquad (\text{IV.2})$$

The probability that a trait is unique to a pair of species (A and B, for example) is

$$\frac{N_{A \cap B} - N_{A \cap B \cap C}}{N_{A \cup B \cup C}} \qquad (\text{IV.3})$$

and the probability that a trait is shared by all three species is

$$\frac{N_{A \cap B \cap C}}{N_{A \cup B \cup C}}. \qquad (\text{IV.4})$$

Although we do have pairwise "distances" from which the two-way intersections might be inferred, we have no data regarding the three-way intersection ($N_{A \cap B \cap C}$). The pairwise distances seem to indicate little about this three-way intersection. For problems involving more than three species, additional unknown values would emerge involving several three-way intersections, four-way intersections, etc. It is thus possible that an entire group of species could be weighted minimally because they are redundant with each other, regardless of the *group's* uniqueness relative to the other species.

With regrets, then, we do not treat genetic diversity further here. We abandon this topic knowing that important diversity information will be lost if genetic diversity is ignored. When focusing on a species richness objective, we may end up emphasizing a group of species that represents less genetic diversity than a smaller group with more genetic variability. This problem could probably be mitigated in practice by considering not only species richness and habitat diversity, but also populations of threatened or endangered species that may require specific management actions, management of specialized or limited habitats, maintenance of a variety of ecosystem functions, and other special considerations.

In chapters 12 and 13, we discuss an approach that optimizes the distribution of diverse habitat types so as to maximize several measures of species richness. In this context, it addresses both habitat diversity and species richness. These two measures of diversity are interrelated because different wildlife species favor different types of habitat and the optimal combination is not obvious. NcNeely et al. (1990) and others have identified habitat protection as a principal means of providing for diverse species, subspecies, or population gene pools. Thomas and Salwasser (1989) and Westman (1990) have pointed out that with limited available lands, seminatural and managed habitats as well as conservation preserves must contribute to meeting biodiversity management goals. Their view is consistent with legislation and evolving policy for public forest lands in the United States (Wilkinson and Anderson, 1987). Chapter 12 develops the objective functions and demonstrates their application to the model discussed in chapter 8. Chapter 13 then addresses the important topic of the sustainability of species richness objectives.

Chapter 12

SPECIES RICHNESS OBJECTIVE FUNCTIONS

In order to link management actions to species richness, we assert that the structure of the ecosystem is affected by management actions that change vegetative cover, and that this structure affects wildlife populations. In this setting, it is typically not obvious what distribution of habitat structures is optimal because different wildlife species favor different types (for example, forest stand ages) and the species richness objective function is not clear. It is presumed that viability of wildlife species is the underlying objective, and that the probability of viability is a function of species population size. The problem addressed in this chapter is to define objective functions that measure the species richness of terrestrial wildlife (see Hof and Raphael, 1993). We begin by discussing the determination of an optimal steady state without concern for management action allocation across time or space. We then address the dynamic spatial allocation problem with the model developed in chapter 8 (see also, Hof et al., 1994). Some steady states, such as uniform middle-

Much of this chapter was adapted from: J. G. Hof and M. G. Raphael, Some mathematical programming approaches for optimizing timber age class distributions to meet multispecies wildlife population objectives, *Canadian Journal of Forest Research* 23 (1993): 828–834, with permission from the publisher, the National Research Council of Canada Research Press; and J. Hof, M. Bevers, L. Joyce, and B. Kent, An integer programming approach for spatially and temporally optimizing wildlife populations, *Forest Science* 40(1) (1994): 177–191, with permission from the publisher, the Society of American Foresters. The dynamic objective functions presented here were developed in Bevers, Hof, Kent, and Raphael (1995) and are further examined in chapter 13.

aged timber, would be very difficult (or impossible) to maintain over time. In this case, the steady state is not sustainable, and the reader is referred to the discussion of dynamic equilibria in chapter 13.

Determining the Optimal Steady State

The choice variables we use to determine the steady state are the number of hectares in different age classes of timber, which are regarded as different types of habitat. The following formulation assumes that species populations are a linear function of these choice variables. Spatial configuration of the age classes is addressed later in the chapter. Beyond spatial independence, the linearity assumption implies that estimates of relative abundance of a species are independent of abundance of all other species. We know that predator–prey and other species interactions violate this assumption. Our approach may yield misleading estimates if one member of an interacting pair of species is differentially affected by changing the allocation of a particular habitat while the other species is not. For example, if an optimum allocation results in a greater abundance of a predator, its prey populations may be depressed below the level predicted by these models.

The expected number of animals in each species population is determined by

$$S_i = \sum_{j=1}^{m} a_{ij} X_j \qquad \forall\, i, \qquad (12.1)$$

$$\sum_{j=1}^{m} X_j \leq A, \qquad (12.2)$$

where
X_j = the number of hectares allocated to the jth age class,
a_{ij} = the number of animals of species i per hectare of land allocated to the jth age class,
S_i = the expected number of animals in the species i population,
n = the number of animal species considered,
m = the number of age classes, and
A = the number of hectares available.

It will also be useful to define T_i as

$$T_i = (\text{MAX}\, a_{ij}) \cdot A \qquad \forall\, i, \qquad (12.3)$$

where MAX a_{ij} is the maximum a_{ij} across all j for each i.

Thus T_i is the maximum number of animals in the species i population that can be obtained from the A hectares. It is assumed that the probability of a species being viable in the study area over a given period of time (V_i) is a function of the number of animals of that species (see Marcot et al., 1986; Mace and Lande, 1991) and that if T_i animals are present, the probability of viability is near 1. Two functional forms for the V_i are considered: linear and logistic. The advantage with the first is that it allows optimization with linear programming. The second requires nonlinear programming for absolute precision, but is more realistic. Piecewise approximation of the logistic form is also used, and is discussed in the Appendix for this chapter. The logistic form indicates a slow increase in viability at low population size, an accelerated rate at intermediate populations, and a slow rate at high population sizes (figure 12.1).

The linear V_i are defined as

$$V_i = \frac{1}{T_i} \cdot S_i \qquad \forall\, i, \tag{12.4}$$

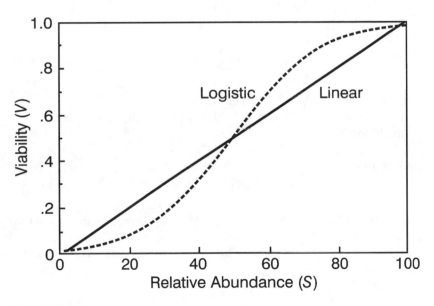

FIGURE 12.1
Comparison of linear and logistic relationships between population viability (probability) and relative abundance (scaled between 0 and 100, where 100 is maximum population size).

where V_i is the probability of viability. These linear V_i functions have the properties that when $S_i = T_i$, $V_i = 1$; when $S_i = T_i/2$, $V_i = .50$; and when $S_i = 0$, $V_i = 0$.

The logistic V_i are defined as

$$V_i = \frac{1}{1 + \exp\{[(T_i/2) - S_i]/(T_i/8)\}}. \tag{12.5}$$

These logistic V_i functions have the properties that when $S_i = T_i$, $V_i = .982$; when $S_i = T_i/2$, $V_i = .50$; and when $S_i = 0$, $V_i = .018$. We recognize that it is illogical to indicate a .018 probability of viability with 0 population, but the standard logistic function is asymptotic to 0. If this presents a problem in practice, a constant value (.018) could be subtracted from all V_i without affecting the optimal solution.

We present three approaches to optimizing this problem: maximize the expected value of the number of viable species (MAXSUM), maximize the minimum probability of viability across species (MAXMIN), and maximize the joint probability of all species being viable (MAXJPR). To maximize the expected value of the number of viable species, we do the following:

Maximize

$$\sum_{i=1}^{n} V_i, \tag{12.6}$$

subject to (12.1), (12.2), and either (12.4) or (12.5).

To maximize the minimum probability of viability across species, we do the following:

Maximize λ,

subject to

$$\lambda \leq V_i \qquad \forall i \tag{12.7}$$

and (12.1), (12.2), and either (12.4) or (12.5), where λ is the minimum probability of viability across species.

To maximize the joint probability of all species being viable, we do the following:

Maximize

$$\prod_{i=1}^{n} V_i, \tag{12.8}$$

subject to (12.1), (12.2), and either (12.4) or (12.5).

The first approach is the most efficiency-oriented in terms of maintaining the highest possible number of species. The second approach maximizes evenness of abundance because the poorest probability of viability is maximized. The third approach is somewhat of a middle ground because some species are allowed to have low viability probabilities, but this is penalized when the joint probability of all species being viable is maximized; the entire objective function value becomes 0 if the viability probability of any one species is 0. We treat all species equally with respect to priority for conservation. This fits legal mandates such as the National Forest Management Act of 1976, which applies to lands managed by the USDA Forest Service and directs national forests to manage habitats for all existing native and desired nonnative plants, fish, and wildlife species to maintain at least viable populations. Other authors (such as Reid and Miller, 1989) have suggested that so-called weedy species associated with disturbed habitats should be given less weight in diversity analyses. Raphael (1988) and Raphael et al. (1988) point out that many early-seral species are abundant and widely distributed, whereas many late-seral species have restricted ranges and narrow ecological tolerance, which might justify preferential treatment in some cases.

Population viability, the likelihood that a population will persist with a given probability for a given time period (Shaffer, 1981), is a complex concept and full evaluation of population viability would require demographic data far beyond simple abundance measures (Soulé, 1987). Our approach is probably most useful to identify species that are at risk because of low abundance. Further study could then be initiated to develop a more comprehensive viability assessment for those species. Such considerations suggest that equity across species must be considered along with general efficiency measures.

A Steady-State Example

Our example involves an area of about 1.1 million ha of Douglas-fir (*Pseudotsuga menziesii*) forest in northwestern California. Estimates of relative abundance of 92 terrestrial vertebrates in five timber age classes (the same classes used in chapter 8) were obtained from recent studies of breeding birds, reptiles, amphibians, and mammals (Raphael, 1988; Raphael et al., 1988). The age classes were early brush/sapling (ESS, regenerating stands less than 10 years old), late brush/sapling (LSS, 10–20 years old), young forest (YNG, 20–150 years old), mature forest (MAT, 150–250 years old), and old growth forest (OLD, more than 250 years old). Detailed structural characteristics of the age classes are given by Marcot (1984) and Raphael (1988).

The first two approaches (MAXSUM and MAXMIN) with linear V_i functions are solvable as linear programs. The third approach (MAXJPR) with linear V_i functions is a nonlinear program, and all three approaches with logistic V_i functions are nonlinear programs. To avoid very small numbers in the solution procedure (the product of 92 probabilities), the V_i in the MAXJPR approach had to be scaled between 0 and 3 for the linear V_i and 0 and 3.5 for the logistic V_i. The results reported have been scaled back to the actual joint probabilities, which are indeed very small numbers. An alternative approach is the Miller and Wagner (1965) transformation, which uses the log likelihood function. Instead of the product of probabilities, the sum of the logs of the probabilities is maximized. This approach also yields a convex program, and it produced identical results to the scaling approach (as one would expect). In the nonlinear programs for the MAXSUM approach with logistic V_i, convexity is not ensured, so local optima are possible. For these cases, a variety of starting points were tested, and the solutions presented appear to be global optima on that basis.

Results

The three objective functions resulted in different estimates of optimum age class distribution (figure 12.2). Estimates using the MAXMIN and MAXJPR approaches and based on linear V_i produced fairly equitable allocations of area across age classes. Conversely, the linear MAXSUM approach allocated 100 percent of the area to young forest. This approach simply allocates the entire area to the age class that has the highest total relative (to the maximum for each species) population. It treats all units of viability uniformly, such that, for example, the first .1 probability of viability for one species is valued the same as a change from .9 to 1.0 for another species. The logistic MAXMIN approach resulted in an identical allocation to the linear MAXMIN model. This result was expected for the MAXMIN approach because the V_i are identical, monotonically increasing functions of population (see Hof and Pickens, 1991). In fact, the MAXMIN objective functions could just as easily be based on standardized population variables (such as S_i/T_i) with identical results. This simply reflects our assertion that probability of viability increases identically for each species as a function of relative population. Thus maximizing the minimum relative population is the same as maximizing the minimum probability of viability as defined here. The logistic MAXSUM approach

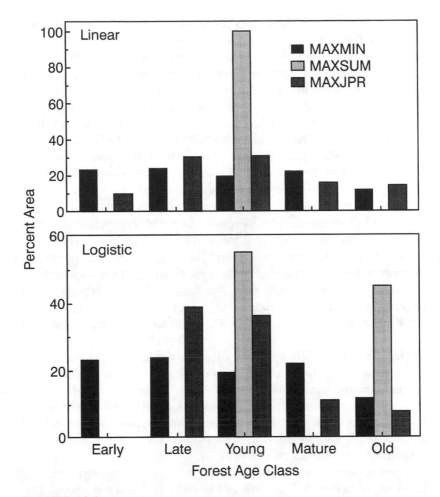

FIGURE 12.2

Optimum allocations of age classes of Douglas-fir forest calculated using three approaches assuming either linear or logistic relationships between viability and abundance. Age classes are early regenerating stands less than 10 years old; late regenerating stands 10–20 years old; young forest 20–150 years old; mature forest 150–250 years old; and old growth forest more than 250 years old.

allocated 55 percent of the area to young forest and all remaining area to old growth forest; the logistic MAXJPR allocated most of the area to late brush and young forest and much less to mature and old growth. Most of these solutions do not appear to be sustainable over time. Chapter 13 addresses this sustainability question in some detail.

Land areas allocated by each approach resulted in distinct distributions of the V_i (figure 12.3). Linear MAXMIN yielded a minimum V_i of .24. Minimum V_i of the logistic MAXMIN was .11. The MAXSUM approach with both linear and logistic viability functions created highly inequitable results across species, with modes at very low (below .1) and very high (above .9) probabilities. Distribution of the V_i based on the linear MAXJPR model was approximately normal, with mean .46 and median .44. Distribution of the logistic MAXJPR V_i values was similar but with slightly more species at low and high probabilities. The results are sensitive to the choice of viability functional form, and the logistic function is more realistic. Piecewise approximation of the logistic function is addressed in the Appendix, and used in the next section of this chapter, which includes spatial considerations.

Of the three approaches, MAXMIN and MAXJPR seem to best meet objectives based on diversifying risk of extinction across the various species. The MAXMIN approach directly spreads out the uncertainty across species and the MAXJPR penalizes low probabilities in calculating the joint probability as a product. These approaches are preferable when it is desired to equitably distribute risk across species. The MAXSUM approach would be appealing only when species equity is not important. As discussed earlier, equity would be important in most potential applications of this type of model.

It was assumed that all species had symmetrical viability probability functions, such that the probability of viability is .5 when the population is one-half the maximum (T_i). Table 12.1 presents the results from one relaxation of that assumption. Viability probability functions were skewed to the left for the birds and to the right for the other animals (implying for illustration purposes that the probability of viability is much more sensitive to population for the other animals than for the birds). The modified functions for the birds are

$$V_i = \frac{1}{1 + \exp\{[(T_i/4) - S_i]/(T_i/16)\}}. \qquad (12.9)$$

These have the properties that when $S_i = T_i$, $V_i = .9999$; when $S_i = T_i/2$, $V_i = .982$; when $S_i = T_i/4$, $V_i = .50$; and when $S_i = 0$, $V_i = .018$. The

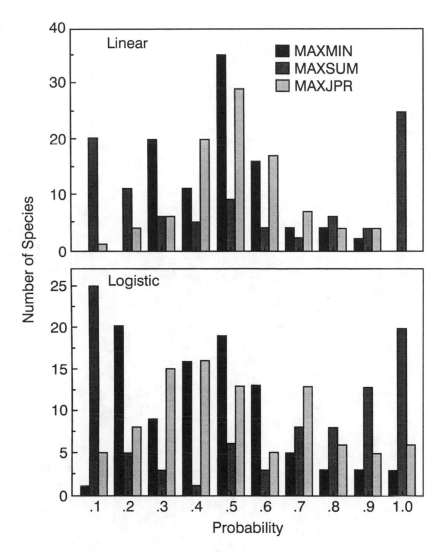

FIGURE 12.3

Distribution of probabilities of viability (V_i) across 92 terrestrial vertebrate species under optimum allocations of age classes of Douglas-fir forest (figure 12.2) for three approaches assuming either linear or logistic relationships between viability and abundance.

TABLE 12.1

A Comparison of Model Results with Symmetrical (and similar) and Skewed (and dissimilar) Logistic Viability Probability Functions

	MAXMIN		MAXSUM		MAXJPR	
	Symmetrical	Skewed	Symmetrical	Skewed	Symmetrical	Skewed
Hectare allocation to age classes (100s)						
Early brush/sapling	2525	3069	0	0	0	1583
Late brush/sapling	2603	2147	0	4373	4242	3092
Young	2111	2298	6009	3694	3961	2622
Mature	2384	742	0	455	1236	1903
Old growth	1278	2645	4891	2377	1460	1700
Objective function value	0.110	0.069	47.6	55.7	6.14×10^{-41}	9.51×10^{-37}

modified functions for the other animals are

$$V_i = \frac{1}{1 + \exp\left[(T_i - S_i)/(T_i/4)\right]}. \qquad (12.10)$$

These have the properties that when $S_i = 2T_i$, $V_i = .982$; when $S_i = T_i$, $V_i = .50$; and when $S_i = 0$, $V_i = .018$. The results in table 12.1 show that model solutions are sensitive to the viability probability function, even with a given functional form. The allocation of hectares to age classes and objective function values are quite different. Most notably, the skewing of the viability probability functions evens out the land allocation for the MAXSUM and MAXJPR approaches. This is not a general result, but indicates that the model results are sensitive to model assumptions and parameter specifications.

Allocation over Time and Space

In order to demonstrate how the objective functions just analyzed can be applied to a spatial scheduling model, we apply them to the model presented in chapter 8. We take the joint probability of all species being viable as the species richness metric for a given time period. Because the model in chapter 8 is dynamic, a metric across time periods is needed. We discuss this topic in more detail in chapter 13, in the context of sustainability of species richness. For now, we simply demonstrate two dynamic objective functions: a MAXMIN: J_t approach that maximizes the minimum joint species probability across time periods and a MAXJV approach that maximizes the joint viability of all species across all time periods. As noted earlier, maximizing the joint probability in any time period is equivalent to maximizing the sum of the natural logs of the viability functions (see Miller and Wagner, 1965; Hof and Pickens, 1991). Miller and Wagner (1965) show that the log transformation is concave (leading to a convex program) for many common probability density functions, including the normal. In the Appendix, we show that the sum of the logs of the logistic viability functions is also concave, so maximization with linear methods using piecewise approximation is tenable. The actual joint probability is determined by taking the antilog after solution. The MAXMIN: J_t approach would thus add the following to equations (8.1)–(8.5) in chapter 8:

Maximize λ, subject to

$$\lambda \leq \sum_i \sum_j l_j p_{itj} \qquad \forall\, t, \qquad (12.11)$$

$$p_{itj} \le b_j \qquad \forall\, j \qquad\qquad (12.12)$$
$$\forall\, i$$
$$\forall\, t,$$

$$F_{it} = \sum_j p_{itj} \qquad \forall\, i \qquad\qquad (12.13)$$
$$\forall\, t.$$

Note that (12.11) approximates

$$\lambda \le \prod_i V_{it} \qquad \forall\, t, \qquad\qquad (12.14)$$

$$V_{it} = v(F_{it}) \qquad \forall\, i \qquad\qquad (12.15)$$
$$\forall\, t.$$

Thus

$$\ln \prod_i V_{it} = \sum_i \ln\big[v(F_{it})\big],$$
$$\qquad\qquad\qquad\qquad\qquad (12.16)$$
$$\ln \prod_i V_{it} \approx \sum_i \sum_j l_j p_{itj},$$

where

j indexes approximation segments,

v = a logistic viability function,

p_{itj} = the jth segment of the total population (F_{it}) in time period t for species i,

l_j = a coefficient that approximates $\ln(v)$ for the jth segment,

b_j = the upper bound on the size of the jth segment (p_{itj}), which is the same for all i and t, and

V_{it} = the probability of viability for species i in time period t.

Equation (12.11) defines λ as the minimum periodic joint probability of viability across time periods. Equations (12.12) and (12.13) define the segments, p_{itj}, for piecewise approximating the natural log of the logistic viability function. Each l_j is calculated as the slope of the line approximating $\ln[v(F_{it})]$ for the associated p_{itj} segment. When λ is maximized, the minimum of the approximated $\ln \prod_i V_{it}$ is maximized. In chapter 13, we include further analysis for the case where some of the equations in (12.11) are not binding.

The joint probability of all species being viable in all time periods (MAXJV) could be optimized (approximately) by retaining (12.12) and (12.13), but replacing the objective function and (12.11) with the following:

Maximize

$$\sum_t \sum_i \sum_j l_j p_{itj}. \qquad (12.17)$$

In both approaches, it is necessary to take the antilog of the objective function solution value to obtain the approximated joint probability.

For demonstration with the model from chapter 8, the logistic viability functions in equation (12.15) are defined for species i and time period t as

$$V_{it} = \frac{1}{1 + \exp\{[(100/2) - F_{it}]/(100/8)\}} \qquad (12.18)$$

These logistic functions have the properties that when $F_{it} = 100$, $V_{it} = 0.982$; when $F_{it} = 50$, $V_{it} = 0.5$; and when $F_{it} = 0$, $V_{it} = 0.018$.

Results

Figures 12.4 and 12.5 present solutions that use the piecewise-approximated logistic viability objective functions combined with the constraint set in equations (8.1)–(8.5) in chapter 8. Figure 12.4 applies the MAXMIN: J_t approach. The result (table 12.2) is a fairly balanced population structure between both species and time periods. In figure 12.4, the spatial layout includes a cluster of cells that are harvested in time period 1 to immediately balance the species 1 population with the species 2 population, and then a smattering of harvests across space and time to

TABLE 12.2
Solutions Values for Figures 12.4 and 12.5

	Figure 12.4	Figure 12.5
Objective function	.2346	.01376
Species 1 population		
$t = 1$	25.00	25.00
$t = 2$	45.40	45.40
$t = 3$	53.70	48.00
$t = 4$	53.00	51.00
Species 2 population		
$t = 1$	75.00	75.00
$t = 2$	54.00	54.00
$t = 3$	46.00	52.00
$t = 4$	47.00	49.00

2	3	2	3	0
2	1	1	0	0
3	1	1	1	3
2	1	1	3	0
3	2	0	0	0

FIGURE 12.4
Time periods of cell harvest (0 = no harvest) when the minimum periodic joint probability of both species being viable across three time periods is maximized.

0	3	3	0	0
0	1	1	3	0
2	1	1	1	0
2	3	1	1	3
3	2	3	0	0

FIGURE 12.5
Time periods of cell harvest (0 = no harvest) when the joint probability of both species being viable in all time periods is maximized.

maintain the balance of species populations in later time periods. Absolute optimality could not be ensured for this model (as with the MAXMIN approach in chapter 8), but the solution given is within 0.4 percent of optimality in terms of objective function attainment, based on the solution bound as discussed in chapter 8.

The solution depicted in figure 12.5 uses the MAXJV approach. In calculating the joint probability as a product of individual viability probabilities, lower probabilities are heavily penalized, so this approach has results that are not very different from the MAXMIN: J_t approach results in figure 12.4. The populations are again fairly balanced across species and time periods (table 12.2). The spatial layout in figure 12.5 again includes a cluster of cells that are harvested in time period 1 to immediately balance the species 1 population with the species 2 population, followed by a smattering of harvests across space and time to maintain the balance of species populations.

It is obvious that two species and three time periods are inadequate to address the questions about species richness sustainability raised earlier in the chapter. In chapter 13, we forgo spatial considerations in order to address these questions.

Appendix

Miller and Wagner (1965) explored concavity of natural logarithm transformations of probability distributions for establishing global optimality in chance-constrained nonlinear programming problems. They identified a large and important class of distributions, which includes the normal family (Hof and Pickens, 1991), that are rendered concave by the logarithm transformation. Linear programming necessitates piecewise linear approximation of these and other nonlinear objective functions. This approximation involves breaking the domain of the independent variable into some number of discrete pieces. Each piece is then represented in the linear programming matrix as a new, bounded variable such that the sum of the activities for these variables equals the total activity for the original independent variable. Linear approximations of the slope of the dependent variable curve are then estimated for each segment. Positive linear combinations of concave (convex) functions (that is, where all coefficients are positive real numbers) are themselves concave (convex). Thus such combinations of concave (convex) functions that have been piecewise approximated provide suitable maximization (minimization) objective functions for linear programs. Concavity (convexity) requirements for global maximization (minimization) with linear methods imply that the per unit values of these piecewise approximations are monotonically decreasing (increasing). We briefly demonstrate that piecewise linear approximation of natural logarithm transformations of smoothly continuous single-variable functions are monotonically decreasing when the ratio of the derivative of the original function to the function itself is monotonically decreasing. We then show that this is true for the general logistic distribution.

Declining Monotonicity of Natural Logarithm Transformations

Let f be a smoothly continuous, nonlinear function of the continuous variable X. Because

$$D_x \ln f(x) = f'(x)/f(x), \quad f(x) > 0, \quad (12.19)$$

(Chiang, 1984) we can see that the slope of the transformed function is monotonically decreasing only when the ratio of the derivative of the untransformed function to that function is monotonically decreasing.

To see that this is also true for linear approximation between two points, x_1 and x_2, let $x_2 = x_1 + \Delta x$, with $x_1 > 0$ and some constant $\Delta x > 0$. Then, between x_1 and x_2,

$$D_x \ln f(x) \cong \frac{\ln f(x_2) - \ln f(x_1)}{\Delta x} = g(x_1) \quad (12.20)$$

because x_2 is itself an incremental function of x_1. In order for this approximation to be monotonically decreasing, the derivative of $g(x_1)$ must remain less than 0. Define $h(x_1) = f(x_2(x_1))$. Then,

$$g(x_1) = \frac{1}{\Delta x} \ln h(x_1) - \frac{1}{\Delta x} \ln f(x_1) \quad (12.21)$$

and

$$
\begin{aligned}
D_{x_1} g &= \frac{1}{\Delta x} \frac{D_{x_1} h}{h} - \frac{1}{\Delta x} \frac{D_{x_1} f}{f} \quad (12.22) \\
&= \frac{1}{\Delta x} \frac{D_f h \cdot D_{x_2} f \cdot D_{x_1} x_2}{h} - \frac{1}{\Delta x} \frac{D_{x_1} f}{f} \\
&= \frac{1}{\Delta x} \frac{D_{x_2} f(x_2)}{f(x_2)} - \frac{1}{\Delta x} \frac{D_{x_1} f(x_1)}{f(x_1)}.
\end{aligned}
$$

Keeping this slope less than 0 implies

$$\frac{1}{\Delta x} \frac{D_{x_1} f(x_1)}{f(x_1)} > \frac{1}{\Delta x} \frac{D_{x_2} f(x_2)}{f(x_2)} \tag{12.23}$$

or

$$\frac{f'(x_1)}{f(x_1)} > \frac{f'(x_2)}{f(x_2)} \qquad \text{Q. E. D.} \tag{12.24}$$

We also note that the choice of step size, Δx, is of no significance except for determining the number of piecewise variables required to cover the domain of the independent variable. Generally, the approximation improves with smaller choices of step size.

The Logistic Distribution

To establish that such approximations are suitable for the natural logarithm transformation of the logistic distribution, we must show that the ratio of the derivative of the logistic cumulative density function (CDF) to that CDF is monotonically decreasing. This is the ratio of the probability density function (PDF) to the CDF. These functions are given by Johnson and Kotz (1970) as

$$f_X(x) = \frac{\exp\{(\alpha - x)/\beta\}}{\beta[1 + \exp\{(\alpha - x)/\beta\}]^2}, \qquad \beta > 0, \tag{12.25}$$

and

$$F_X(x) = \frac{1}{1 + \exp\{(\alpha - x)/\beta\}}, \qquad \beta > 0. \tag{12.26}$$

The ratio of the two simplifies to

$$\frac{f}{F} = \frac{1}{\beta} \cdot \frac{\exp\{(\alpha - x)/\beta\}}{1 + \exp\{(\alpha - x)/\beta\}}. \tag{12.27}$$

Letting $y = \exp\{(\alpha - x)/\beta\}$, we note that as x increases, y decreases. Because $\beta > 0$, we observe that

$$\frac{y}{1 + y} \to 0 \quad \text{as } y \to 0 \quad (\text{and as } x \to \infty). \tag{12.28}$$

The ratio is indeed monotonically decreasing with increasing values of the independent variable x.

Chapter 13

SUSTAINABILITY OF SPECIES RICHNESS

In chapter 12 we examined the problem of allocating habitat types for diverse species to optimize species richness and equity. The results were questionable, however, in terms of sustainability. For example, in a number of cases, static allocation placed lands in young forest seral stages, but not in the earliest seral stage. Consequently, as initially young stands grow old over time those allocations cannot be maintained with ingrowth (maturing of early seral stages into older ones). This suggests a need for evaluating these problems using dynamic models optimized over long time frames. The purpose of this chapter is to explore optimization objective functions that might lead to sustainable forest habitat allocations for a large group of diverse wildlife species (see Bevers et al., 1995). The case study includes the same 92 species of wildlife investigated in chapter 12.

The Modeling Approach

Sustainability has been defined as the achievement and maintenance in perpetuity of continuous or regular periodic levels of resource production (Ford-Robertson, 1971). This suggests conversion from some initial condition to a new state that can be maintained. In this chapter, we as-

Much of this chapter was adapted from M. Bevers, J. Hof, B. Kent, and M. G. Raphael, Sustainable forest management for optimizing multispecies wildlife habitat: A coastal Douglas-fir example, *Natural Resource Modeling* 9 (1995): 1–23, with permission from the publisher, the Rocky Mountain Mathematics Consortium.

sume that the fundamental problem is to convert a given set of forest conditions, through a combination of natural growth and management over time, to a forest system that addresses the joint habitat needs of diverse wildlife species in a sustainable manner. In a tumultuous ecosystem subjected to unexpected disturbances, any long-term equilibrium will probably have shifted before ever being achieved (Kaufman, 1993). Nevertheless, analyzing sustainable habitat allocations in the absence of such unexpected disturbances still provides insights into baseline natural resource planning. Occasional adjustments to account for unexpected events are the motivation for adaptive management (Walters, 1986) and would be handled by successive management plan revisions.

May (1973) and other dynamic modelers have explored equilibrium conditions in simulation models in great detail. In the absence of random or irregular perturbations, three forms of dynamic equilibrium are generally recognized. These include point equilibria, limit cycles, and damped oscillations. A point equilibrium exists when constant conditions are maintained from time step to time step. A limit cycle equilibrium exists when repetitive cycles involving multiple time steps occur. Damped oscillations are also cyclic equilibria, but without indefinite repetition. Instead, these cycles generally tend to converge toward a point equilibrium state. Although it is unlikely that we can rigorously prove that a state of equilibrium exists in a complex dynamic optimization model solution, these are the patterns we might expect to see demonstrated.

We hypothesize that, given a sufficiently long planning horizon, a discrete-time, free-terminal mathematical programming model of forest habitat management for sustainable wildlife diversity will result in identifiable conversion and equilibrium management phases when optimized using objective functions that are oriented toward temporal equity of species viability measures. Three such objective functions are discussed in the next section of this chapter. In order to accommodate the dimensions of the case study, a linear deterministic model is necessary; linear relationships are used wherever possible, and nonlinear functions are piecewise approximated or otherwise transformed into linear functions.

Objective Functions

Population viability is, again, the basic unit of measure for our objective functions. Shaffer (1981) interprets population viability in terms of the likelihood (probability) that a population will persist for a given (usually long) period of time. Specifically, our measure of viability V_{kt} for a given species k in time period t is

$$V_{kt} = \frac{1}{1 + \exp\{[(T_k/2) - N_{kt}]/(T_k/8)\}} \qquad 0 < V_{kt} < 1, \quad (13.1)$$

where N_{kt} is the population of species k in time period t and T_k is the population corresponding to a viability of .982 (see chapter 12). These logistic V_{kt} functions are asymptotic to 0 such that when $N_{kt} = 0$, $V_{kt} \cong .018$. In our approximation of these functions, we impose that viability is 0 if no population is present. The periodic interpretation of species viability represented by V_{kt} is a departure from Shaffer's, but provides a useful way of accounting for managerial adjustments to available habitat over time not typically considered in viability assessments.

For a multispecies dynamic problem, the viability measures, V_{kt}, may be combined into objective functions in any number of ways. It is important to note that when examined probabilistically, Westman's (1990) two diversity components (species richness and equity) provide somewhat competing objectives. Predicted species richness for any future time period may be interpreted as the expected value of the number of species predicted to be present. Under an assumption of species independence, this value would be the sum across all species in a given time period t of V_{kt} (V_{kt} being the expected presence for species k in time period t as a Bernoulli event). Chapter 12 showed that maximizing such an objective (MAXSUM) may not be equitable.

A more equitable approach for a given time period t is joint viability:

$$J_t = \prod_k V_{kt}. \qquad (13.2)$$

With logistic viability functions, maximizing this measure tends to push all species populations toward at least the middle sections of their viability curves. Another approach to equitable habitat allocation in any time period is to maximize the minimum periodic species viability (V_{kt}). This approach provides equity in the manner suggested by Rawls (1971). In both approaches (described in chapter 12), when viability is 0 for any one species, the species richness measure of the entire set of species is 0. Thus these approaches behave usefully only for applications in which all species retain some nonzero probability of viability.

Based on the results from chapter 12, we use individual and joint viability measures within any time period, rather than viability sums. With a dynamic model, the question of how to combine periodic measures over time to encourage temporal equity still remains, whether species equity is of concern or not. We examine three dynamic objective functions. Two are based on the joint measure, J_t, and one is based solely on V_{kt}.

The MAXJV objective function maximizes the product of periodic joint viabilities across time periods:

Maximize

$$\prod_t J_t. \tag{13.3}$$

We form this mathematical product across time periods to encourage temporal equity in much the same manner that the product of within-time-period viabilities (J_t) encourages periodic equity among species. Keeney (1974) describes useful characteristics of such multiplicative utility functions, and Harvey (1985) demonstrates their distributional equity properties. We must be careful, however, not to interpret this mathematical utility function as the joint viability across time periods and species. That would only be true when viability is defined as the probability of a species persisting for a single time period. Because viability is generally estimated for long time spans in order to capture a broad range of stochastic variation in expected survival, such short-term probabilities are not typically considered.

With our second objective function, MAXMIN: J_t, we initially maximize the minimum periodic joint viability as follows:

Maximize λ

subject to

$$\lambda \le J_t \qquad \forall\, t. \tag{13.4}$$

Following this maximization, some subset of these constraints on λ may not be binding (we ignored this possibility in chapter 12). If this is the case, we then repeat the optimization by replacing λ in the previously binding constraints with a fixed value equal to the previous objective function solution value. This sequence of problem revision and maximization is continued until lower bounds have been maximized for all J_t, thus improving joint viability (time period by time period) as much as possible for the entire planning horizon. This iterative optimization method is similar to nucleolus approaches in game theory (Schmeidler, 1969). The MAXMIN: J_t approach is also oriented toward temporal equity because it directly optimizes the time periods that are in the poorest state in terms of joint species viability.

Our third objective function, MAXMIN: V_{kt}, is used to examine equity approaches to management strategies that emphasize single species, such as those involving threatened or endangered species. For this objective, we proceed as follows:

Maximize λ

subject to

$$\lambda \le V_{kt} \qquad \forall\, k,\, t \tag{13.5}$$

iteratively (as before) so that the lowest remaining individual species' periodic viability is maximized at each iteration.

A Coastal Douglas-fir Case Study

The same 92 wildlife species and five forest age classes (ESS, LSS, YNG, MAT, and OLD) defined in chapter 12 are used here. Expected values for populations of the wildlife species vary by forest age class and are used to provide estimates of the carrying capacity for each species in each age class. The largest expected population for a given species, resulting from all forest land falling within the preferred age class of that species, provides an estimate of T_k for estimating viability in equation (13.1).

In this chapter, the shortest time step used to distinguish age classes is 1 decade. Applying even-aged forest management, which is typical for coastal Douglas-fir forests (Williamson, 1973), implies a single age (in decades) for the forest cover on any site in any time period. We defined forest management variables, X_{jm}, to represent the area from initial age class j assigned to treatment schedule m, extending over the entire planning horizon. This approach to defining habitat management options is similar to that used in previous chapters (see also Johnson and Scheurman, 1977).

With age and time discretized into decades, forest age for any area increments by one for each period in which it is unharvested. Harvests reset the cover age to 0 in the period they occur. Based on age, the entire area assigned to a particular management variable falls into one of the five forest seral stages in any time period. Each unit of area in each seral stage results directly in some population carrying capacity for each species. Wildlife carrying capacities are then treated simply as functions of the cumulative forest area in each age class. Consequently, carrying capacity limits can be modeled with linear constraints relating potential wildlife populations in each time period directly to stratified forest management variables according to

$$N_{kt} \leq \sum_j \sum_m c_{jkmt} X_{jm} \qquad \forall\, k,\, t, \qquad (13.6)$$

with initial age stratification constraints

$$\sum_m X_{jm} = A_j \qquad \forall\, j, \qquad (13.7)$$

where N_{kt} is the population of species k in time period t and c_{jkmt} is the expected carrying capacity (per unit area) of species k in time period t if forest management prescription m is implemented for initial age stratum j. A_j is the area of each initial forest age stratum.

To limit wildlife species' population growth, we use simple exponential function r values, r_k, that indicate a growth potential per decade in the absence of other limiting factors for each species. These constraints are

$$N_{kt} \leq (1 + r_k)N_{k,t-1} \qquad \forall\, k \qquad (13.8)$$
$$t = 2, \ldots, P$$

If initial populations were known, the carrying capacity constraints of equation (13.6) for the first time period could be replaced by fixed activity levels for the N_{k1} variables. In the absence of this data, we set the initial wildlife populations at the carrying capacities determined by the initial forest conditions. For all time periods after the first one, N_{kt} is determined by whichever of (13.6) or (13.8) is limiting.

Although this model is a simplified view of a complex and highly interactive forest wildlife community, it does capture the basic potentially limiting dynamic factors of population growth and carrying capacity. The Appendix to this chapter describes the relationship between this model and a conventional systems ecology formulation, documenting the simplifying assumptions. Exponential population growth and no-growth-above-carrying-capacity constraints are two extremes of population dynamics. Many potentially influential factors that could result in intermediate rates of population growth are not considered. Model size and solvability limitations prevented inclusion of additional dynamics, such as spatial population dispersal (as in chapter 8) or population-dependent growth. Even so, the model is quite suitable for examining the use of temporal-equity–oriented objective functions in optimizing dynamic equilibria.

Linear Approximation of Objective Functions

Both the MAXJV and the MAXMIN: J_t objective functions are composed entirely of products (J_t) of the periodic wildlife species viabilities (V_{kt}). As discussed in chapter 12, identical habitat allocations can be achieved by maximizing the sum of natural logarithms (log likelihood) of V_{kt}. This sum can be approximated in linear programming maximization problems because the slopes of piecewise linear approximations of the $\ln(V_{kt})$ viability functions are monotonically decreasing with increasing populations N_{kt} (see the Appendix to Chapter 12).

The linear approximation of the MAXJV objective function from equation (13.3) is as follows:

Maximize

$$\sum_k \sum_l \sum_t v_{kl} n_{klt}, \tag{13.9}$$

subject to

$$n_{klt} \le b_{kl} \qquad \forall \, k, \, l, \, t,$$
$$N_{kt} = \sum_l n_{klt} \qquad \forall \, k, \, t, \tag{13.10}$$

where

l indexes linear approximation segments,

V_{kl} = a coefficient that approximates the ln-logistic function of segment l for each species k,

n_{klt} = the lth segment of the total population (N_{kt}) for species k in time period t, and

b_{kl} = the upper bound on the size of the lth segment for each species, identical for all time periods.

The linear approximation of the MAXMIN: J_t objective function from equation (13.4) is as follows:

Maximize λ

subject to

$$\lambda \le \sum_k \sum_l v_{kl} n_{klt} \qquad \forall \, t, \tag{13.11}$$

$$n_{klt} \le b_{kl} \qquad \forall \, k, \, l, \, t,$$

$$N_{kt} = \sum_l n_{klt} \qquad \forall \, k, \, t.$$

For the MAXMIN: V_{kt} objective in (13.5), we note that each V_{kt} is a monotonically increasing function of the respective N_{kt} variable. Furthermore, all V_{kt} values range identically between 0 and 1. Identical allocations can therefore be achieved as follows:

Maximize λ

subject to

$$\lambda \le \mathbf{N}_{kt} \qquad \forall \, k, \, t, \tag{13.12}$$

where $\mathbf{N}_{kt} = N_{kt}/T_k$ is a standardized population for each species in each time period that results from rescaling the c_{jkmt} carrying capacity coefficients of (13.6).

The case study model consists of equations (13.6)–(13.8) as dynamic process constraints, combined with either (13.9)–(13.10), (13.11), or

(13.12) as the objective function. For evaluating model reduction alternatives (in the following section), joint viability in the last time period is maximized as an analog to static optimization.

Model Reduction

The need for long planning horizons results in very large model sizes and creates solution difficulties, even with linear functions. Two parameters affecting model size, the number of ln-viability function linear approximation segments and the number of species or representative species groups, are explored using a 20-decade analysis with the forest made up initially of old growth (OLD). J_{20} is maximized, first varying the number of approximation segments, then the number of representative species groups. The model is not very sensitive to the number of linear approximation segments used. To examine sensitivity to use of species groups, species are clustered on the basis of habitat preferences using absolute distance in a clustering algorithm. Each group is represented in the analysis by an average habitat preference, and the objective function is weighted according to the number of species in that group. Although the model is somewhat more sensitive to grouping species than to reductions in the number of approximation segments, the effects are still small.

Based on the results of these sensitivity analyses, we elected to use 30 ln-viability linear approximation segments and 40 species groups for the dynamic analyses discussed in the following sections of this chapter. With this approximation, the overall deviation in allocation in decade 20 from the equivalent nonlinear model (from chapter 12) is less than 4 percent of the land base.

Sensitivity to Minimum Harvest Age

Minimum harvest age (MHA) affects the dynamic flexibility included in the model because it controls the set of X_{jm} management variables available for selection. The greatest management flexibility occurs when variables are included with the MHA equal to one time period (1 decade in this case). This means that areas can be treated as often as every decade, allowing those lands to be kept in any sustainable age class distribution. We examine the effect of minimum harvest age by varying MHA from 1 to 7 decades using both the MAXJV and MAXMIN: J_t objective functions and a 20-decade planning horizon. We set initial conditions by stratifying the forest into four beginning age classes made up of 40 percent LSS (age = 1 decade), 38 percent YNG (age = 8), 8 percent MAT (age = 20), and 14 percent OLD (age = 25). In all cases, X_{jm} variables involving treatments for all time periods at or beyond MHA (as well as no-harvest variables) are included.

Tables 13.1 and 13.2 show the allocation and objective function results (scaled) with MHA = 1 for the MAXJV and MAXMIN: J_t objectives, respectively. Three phases are present in each table, although the differences between phases are somewhat subtle in both. The MAXJV solution in table 13.1 shows a distinct conversion phase, through which the allocations fluctuate substantially, from decades 1 through 7. A dynamic equilibrium phase appears from decade 8 through about decade 14. During those periods, the YNG, MAT, and OLD age classes remain in point equilibria with almost no fluctuation. The ESS and LSS age

TABLE 13.1

Optimal Age Class Allocations and Objective Function Contributions over 20 Decades for the MAXJV Objective with a Minimum Harvest Age of 1 Decade

		Percentage of Area Allocated				Joint Viability
Decade	ESS	LSS	YNG	MAT	OLD	(scaled log likelihood)
1	16.87	26.90	34.23	8.00	14.00	328.72
2	22.34	16.87	38.80	8.00	14.00	325.76
3	19.23	22.34	36.43	8.00	14.00	327.69
4	20.96	19.23	37.81	8.00	14.00	326.69
5	19.94	20.96	37.10	8.00	14.00	327.29
6	20.01	19.94	38.05	0	22.00	327.04
7	19.94	20.01	38.05	0	22.00	327.07
8	19.66	19.94	33.21	8.91	18.28	327.42
9	19.94	19.66	33.22	8.90	18.28	327.30
10	19.66	19.94	33.22	8.90	18.28	327.42
11	19.94	19.66	33.22	8.90	18.28	327.30
12	19.66	19.94	33.22	8.90	18.28	327.42
13	19.94	19.66	33.22	8.90	18.28	327.30
14	19.66	19.94	33.25	8.90	18.26	327.42
15	20.04	19.66	32.86	9.18	18.26	327.28
16	19.55	20.04	33.33	8.88	18.20	327.46
17	20.53	19.55	32.50	9.22	18.20	327.16
18	18.50	20.53	34.31	8.33	18.33	327.74
19	22.84	18.50	32.35	7.97	18.33	326.35
20	12.73	22.84	35.87	10.22	18.33	328.84
						$\sum$ = 6546.67

TABLE 13.2

Optimal Age Class Allocations and Objective Function Contributions over 20 Decades for the MAXMIN: J_t Objective with a Minimum Harvest Age of 1 Decade

Decade	ESS	Percentage of Area Allocated				Joint Viability (scaled log likelihood)
		LSS	YNG	MAT	OLD	
1	21.28	23.21	33.52	8.00	14.00	327.13
2	20.76	21.28	35.96	8.00	14.00	327.13
3	20.44	20.76	36.79	8.00	14.00	327.13
4	20.17	20.44	37.39	8.00	14.00	327.13
5	19.86	20.17	37.97	8.00	14.00	327.13
6	19.43	19.86	38.71	0	22.00	327.13
7	18.23	19.43	40.34	0	22.00	327.13
8	18.52	18.23	33.89	8.87	20.49	327.13
9	19.20	18.52	34.13	8.15	20.00	327.13
10	19.20	19.20	34.24	8.09	19.28	327.31
11	19.20	19.20	34.24	8.09	19.28	327.31
12	19.20	19.20	34.24	8.08	19.28	327.31
13	19.20	19.20	34.24	8.08	19.28	327.31
14	19.21	19.20	34.24	8.08	19.27	327.31
15	19.21	19.21	34.24	8.08	19.25	327.31
16	19.23	19.21	34.25	8.08	19.23	327.31
17	19.25	19.23	34.25	8.08	19.19	327.31
18	19.29	19.25	34.26	8.07	19.13	327.31
19	19.35	19.29	34.28	8.06	19.03	327.31
20	16.01	19.35	34.99	10.62	19.03	327.64

$$\sum = 6544.91$$

classes oscillate in small limit cycles that are nearly point equilibria. Beginning with decade 15, the allocations begin to fluctuate more substantially, culminating with a large change in the last period. We associate these fluctuations (particularly in the last period) with a terminal phase in the solution, a phenomenon we examine in a later section on planning horizon length. The MAXMIN: J_t solution in table 13.2 remains in a conversion phase through about decade 9. All age classes enter point equilibrium phases by decade 10, followed by a gradual breakdown into a terminal phase that is clearly distinguishable only in decade 20. Overall, the MHA = 1 solutions differ little between the MAXJV and MAXMIN: J_t objective functions. Both solutions suggest maintaining 19–20 percent of the forest as openings, as shown by the ESS age class.

Increasing minimum harvest age from 1 decade to 2 results in almost no difference in the optimal solution for either objective. This result is consistent with the observation that no harvesting at an age of 1 decade occurs in either MHA = 1 solution, as shown by LSS allocations equaling prior ESS allocations in all periods.

As minimum harvest age is increased beyond MHA = 2, changes in equilibrium allocations result. First, the MAT equilibrium allocation is rapidly reduced (to 0 by MHA = 4) under both objective functions. Overall harvest rates, as evidenced by the ESS allocations, show little change. Instead, greater YNG (and lesser MAT) equilibrium allocations result from the choice of harvest ages. Second, under the MAXMIN: J_t objective, the OLD equilibrium allocation is also reduced (although never quite eliminated in our analysis) by increasing minimum harvest age; by MHA = 7, the YNG equilibrium allocation is more than twice the MHA = 1 YNG allocation shown in table 13.2. This reduction of the OLD allocation does not occur under the MAXJV objective. The allocation changes under the two objectives for MHA = 5 can be seen in tables 13.3 and 13.4.

Another change in results produced by increasing MHA is a definite shift from point equilibria to cycles under both objective functions. Limit cycles appear that are equal in cycle length to the minimum harvest age beginning with MHA = 5. The MHA = 5 cycles (shown by objective function in tables 13.3 and 13.4) are apparent in the ESS, LSS, and YNG age class allocations of decades 10 through 14 and beyond, whereas the OLD age class remains in point equilibrium. The MAT age class has been eliminated, as noted earlier. The correspondence between cycle length and minimum harvest age for MHA = 5, 6, and 7 is demonstrated by the ESS equilibrium allocation cycles in table 13.5. Interestingly, whereas cycle lengths correspond to minimum harvest ages, cycle extremes (smallest and largest values) generally occur in adjacent time periods throughout our analysis.

Increasing MHA also produces increasing differences in the joint viability achieved by the MAXJV and MAXMIN: J_t objective functions. Of course, the log likelihood values decline under both objective functions as management options are reduced by increasing MHA. Nonetheless, both the overall values and the equilibrium phase values are consistently larger under the MAXJV objective function. Conversely, the MAXJV objective function also consistently produces the smallest individual periodic values. In our results, the smallest periodic values for both objective functions always occur during conversion phases, which are present

TABLE 13.3

Optimal Age Class Allocations and Objective Function Contributions over 20 Decades for the MAXJV Objective with a Minimum Harvest Age of 5 Decades

| Decade | ESS | Percentage of Area Allocated | | | | Joint Viability (scaled log likelihood) |
		LSS	YNG	MAT	OLD	
1	0	40.00	38.00	8.00	14.00	331.59
2	17.05	0	60.95	8.00	14.00	309.48
3	11.86	17.05	49.09	8.00	14.00	324.68
4	12.30	11.86	57.05	4.79	14.00	319.49
5	18.14	12.30	50.77	4.79	14.00	322.35
6	15.52	18.14	47.58	0	18.79	325.99
7	17.05	15.52	48.63	0	18.79	324.64
8	16.32	17.05	47.84	0	18.79	325.47
9	14.18	16.32	50.71	0	18.79	324.49
10	18.02	14.18	49.01	0	18.79	323.92
11	15.64	18.02	47.55	0	18.79	325.94
12	17.05	15.64	48.51	0	18.79	324.72
13	16.32	17.05	47.84	0	18.79	325.47
14	14.18	16.32	50.71	0	18.79	324.49
15	18.02	14.18	49.01	0	18.79	323.92
16	15.52	18.02	47.67	0	18.79	325.92
17	17.17	15.52	48.51	0	18.79	324.66
18	16.32	17.17	47.72	0	18.79	325.55
19	19.68	16.32	50.71	0	13.29	324.08
20	18.02	19.68	49.01	0	13.29	325.76

$$\sum = 6482.61$$

in all results. Terminal phases appear in all but one case (the MHA = 5 MAXMIN: J_t solution shown in table 13.4).

By focusing management resources on the most limiting time period or periods the MAXMIN: J_t objective tends to optimize conversion strategy at some expense to long-term levels of viability. Although the loss in overall log likelihood of viability is small when the MAXMIN: J_t objective is used, options in subsequent time periods are still evidently foreclosed to some degree. On the other hand, the MAXJV objective achieves maximum long-term equilibrium levels of joint viability. This achievement comes at some expense to joint viability during the conversion phase. It is clear that the degree to which conversion strategy is

TABLE 13.4

Optimal Age Class Allocations and Objective Function Contributions over 20 Decades for the MAXMIN: J_t Objective with a Minimum Harvest Age of 5 Decades

Decade	ESS	Percentage of Area Allocated LSS	YNG	MAT	OLD	Joint Viability (scaled log likelihood)
1	13.08	40.00	24.92	8.00	14.00	326.38
2	9.97	13.08	54.96	8.00	14.00	320.20
3	17.02	9.97	53.08	5.94	14.00	320.20
4	14.52	17.02	63.04	0	5.42	320.20
5	24.05	14.52	56.01	0	5.42	320.20
6	17.78	24.05	52.76	0	5.42	325.04
7	17.70	17.78	59.10	0	5.42	321.67
8	17.95	17.70	58.93	0	5.42	321.67
9	17.10	17.95	59.53	0	5.42	321.67
10	21.34	17.10	56.14	0	5.42	321.67
11	18.20	21.34	55.04	0	5.42	323.86
12	18.17	18.20	58.21	0	5.42	322.06
13	18.29	18.17	58.12	0	5.42	322.06
14	17.81	18.29	58.47	0	5.42	322.06
15	21.34	17.81	55.43	0	5.42	322.06
16	18.20	21.34	55.04	0	5.42	323.86
17	18.18	18.20	58.21	0	5.42	322.06
18	18.30	18.18	58.12	0	5.41	322.06
19	17.82	18.30	58.47	0	5.41	322.06
20	21.34	17.82	55.43	0	5.41	322.06
						$\sum = 6443.10$

TABLE 13.5

Equilibrium Cycles of ESS Allocation (percentage of forest area) for Minimum Harvest Ages 5–7 Under Two Objective Functions

Decade Within the Cycle	Minimum Harvest Age in Decades 5 MAXMIN: J_t	MAXJV	6 MAXMIN: J_t	MAXJV	7 MAXMIN: J_t	MAXJV
1	21.34	18.02	25.31	19.21	25.29	17.60
2	18.20	15.64	14.42	13.07	12.44	10.55
3	18.17	17.05	15.05	14.88	11.74	13.85
4	18.29	16.32	13.94	11.11	13.05	12.59
5	17.81	14.18	16.50	14.87	10.61	11.63
6			11.83	8.23	15.32	11.77
7					6.65	1.74

slighted under the MAXJV objective is proportional to the number of time periods in the analysis.

Long-term viability of any particular species in an area might be represented by the lowest expected periodic value over the planning horizon. We did not reach an equilibrium with a MAXMIN: V_{kt} objective function across individual species and time periods with our assumed initial conditions. We discuss our results with the MAXMIN: V_{kt} in a later section of this chapter devoted to single-species emphasis.

Sensitivity to Planning Horizon Length

With any dynamic optimization model, a question naturally arises regarding the effect of planning horizon length on solutions, particularly those involving longer times between treatments. We examine this effect by extending the planning horizon to 60 decades and maximizing the MAXJV and MAXMIN: J_t objective functions with a minimum harvest age of 7 decades. Under both objective functions, cyclic equilibrium phases extend throughout most of the planning horizon. The terminal phases (equilibrium breakdown) remain limited to the last few time periods. Table 13.6 shows the resulting cyclic ESS allocations; table 13.7 shows the corresponding objective function levels. In both tables, the MAXMIN: J_t values vary little in response to the longer planning horizon, but the equivalent MAXJV values change more noticeably. The direction of change for the MAXJV solution appears to be toward a more even harvest, allocation, and objective function contribution from period

TABLE 13.6

Optimal Cycles of ESS Allocation over 20- Versus 60-Decade Planning Horizons with a Minimum Harvest Age of 7 Decades

Decade Within the Cycle	Percentage of Area Allocated			
	MAXMIN: J_t		MAXJV	
	20 Decades	60 Decades	20 Decades	60 Decades
1	25.19	25.10	17.60	13.15
2	12.39	12.47	10.55	11.08
3	11.74	11.77	13.85	12.07
4	13.05	13.06	12.59	11.61
5	10.61	10.65	11.63	10.28
6	15.32	15.31	11.77	10.96
7	6.65	6.71	1.74	8.85

TABLE 13.7
Cyclic Joint Viability (scaled log likelihood) Values over 20- Versus 60-Decade
Planning Horizons with a Minimum Harvest Age of 7 Decades

Decade Within the Cycle	MAXMIN: J_t		MAXJV	
	20 Decades	60 Decades	20 Decades	60 Decades
1	313.82	313.86	312.13	318.10
2	325.17	325.07	324.51	321.29
3	313.82	313.86	319.74	319.77
4	313.82	313.86	322.24	320.51
5	313.82	313.86	320.59	319.31
6	313.82	313.86	319.71	318.30
7	313.82	313.86	312.56	317.69

to period as the planning horizon is extended. During the conversion phases, however, the smallest periodic joint viability still occurs (in period 2) under the MAXJV objective function.

Accounting for Mortality

Old growth stands are not harvested during many of the equilibrium phases discussed so far. For example, during the equilibrium phase in table 13.3, all harvests are taken from the YNG age class. The unharvested portion of the MAT age class grows old, after which no additional ingrowth occurs because all YNG stands are harvested before becoming mature. The resulting OLD age class area never changes (until the terminal phase begins). We know, however, that climax coastal Douglas-fir stands are uncommon because of natural stand mortality (Agee, 1991; Hermann and Lavender, 1990). To address this problem, we add deterministic rates of mortality (see Bevers and Kent, 1991) to MHA = 2 and MHA = 7 analyses with a 60-decade planning horizon. Long-term average rates of stand-destroying fire mortality and lags in regenerated stand canopy closure are estimated from national forest planning models in the region. The mortality rates used are 3 percent per decade for stands 2–8 decades old, 2 percent per decade for stands 9–25 decades old, and 1.5 percent per decade for older stands. Seedling/sapling stand canopy closure, or movement from the ESS age class to LSS, is modeled with a 1-decade lag for 20 percent of all new stands to account for regeneration failures caused by seedling mortality.

Essentially identical point equilibria are achieved under both the MAXJV and MAXMIN: J_t objective functions with MHA = 2. The inclusion of mortality rates alters the equilibrium allocations somewhat, but does not alter the point form of equilibrium in the optimal solutions. The equilibrium allocations are ESS = 22.00, LSS = 17.60, YNG = 33.04, MAT = 8.93, and OLD = 18.43 percent of total forest area.

With MHA = 7 and the MAXJV objective function, the mortality rates produce a change in the form of dynamic equilibrium achieved in the optimal solution. Specifically, a damped oscillation converging toward point equilibrium results rather than the limit cycle produced without mortality processes. For example, the ESS allocation converges from cycle extremes of 4.73 and 22.45 percent in periods 6 and 7 to 15.22 and 15.72 percent in periods 55 and 56. These damped oscillations still follow a 7-decade cycle, equal to the minimum harvest age. We could not complete the optimization of the MAXMIN: J_t objective function in this large linear programming matrix because of numeric difficulties accumulated with successive iterations in the solution process.

These mortality rates do not completely correct the problem of mature forest elimination noted in the earlier MHA = 7 solution. Although MAT age class existence is extended through period 41, it is again eliminated thereafter. Imposing an upper limit on the age of old growth stands in the model might be a reasonable approach to this problem, although it seems likely that a longer planning horizon would also be needed to develop equilibrium conditions. It is interesting to note that dynamic models do not necessarily remedy the shortcomings previously noted regarding static equilibria.

Single-Species Emphasis

Both the MAXJV and MAXMIN: J_t solutions for the analysis just described leave one species group at a surprisingly low level of viability, particularly in decades 6 and 7. The group consists of a single species, the white-breasted nuthatch (*Sitta carolinensis*), which has been identified as having very low habitat versatility in other western coniferous forests (Thomas, 1979). It is heavily dependent on mature (MAT) forest stands, with some minor presence in the LSS and OLD age classes. Although we have a mix of seral stages in our initial conditions, many specific ages are initially absent from the YNG, MAT, and OLD age classes. As a result of assigning all young forest (YNG) a starting age of 8 decades and assigning all mature forest (MAT) a starting age of 20 decades, it is impossible for the model to produce any mature forest in decades 6 and 7. Consequently, white-breasted nuthatch viability is reduced to a low of about .0225 in those decades. Even under optimal

MAXJV equilibrium conditions in later decades, white-breasted nut-hatch viability is only about .0443 because of its specialized habitat requirements.

Because the white-breasted nuthatch is found across North America, a habitat manager might be willing to accept 2 decades of locally poor habitat conditions. For purposes of this chapter, however, we use the white-breasted nuthatch to more closely examine the long-term effects of focusing resources on short-term limiting conditions, as the MAXMIN objectives tend to do. Using the same analytical parameters as previously described (MHA = 7 with mortality), we iteratively optimize the MAXMIN: V_{kt} objective function from equation (13.12). White-breasted nuthatch viability is limiting at each iteration. Viability in decades 6 and 7 constrains the initial solution, with decades 13, 14, 18, and 19 constraining the second iteration solution.

The long-term tradeoffs of this approach are noteworthy. With every possible resource focused on nuthatch viability in decades 6 and 7 for the first iteration, the viability in those periods increases only from .0225 to .0233. Unavailability of mature (MAT) forest precludes further improvement in those periods. Because large YNG harvests must be used to produce LSS habitat for even this minor improvement, far greater viability is forgone in later periods by reducing potential MAT allocations. Nuthatch viability remains low for 24 additional decades before again rising toward long-term MAXJV equilibrium levels with the initial iteration solution imposed as lower bounds.

Additional MAXMIN: V_{kt} iterations appear to further exacerbate the problem. As noted earlier, no equilibrium solutions were achieved. It is clear that MAXMIN-type objective functions must be used judiciously with one or a few species. Long-term conditions can be adversely affected by focusing resources on short-term limiting factors. Use of the MAXMIN approach across time periods with broad-based measures such as joint species viability for each time period appears much less likely to yield such extreme results.

A New Definition for a Regulated Forest

The results reported in this chapter suggest that with a species richness objective, the concept of a regulated forest may require some redefinition. Forest regulation has long been studied by foresters as a means of estimating sustainable harvest levels. Clutter et al. (1983) have provided one of the more general definitions by stating that a regulated forest is "any forest that permanently retains a constant age class structure under some constant level of harvest." Note that this definition specifies a constant age class structure, but not any specific age class structure. Even with this general definition, the equilibria results previously reported often occur in cyclic patterns more complex than those traditionally considered for forest regulation.

All three deterministic types of equilibrium are generated as results of optimizing the temporal equity species richness objective functions. The limit cycle and the damped oscillation results clearly depart from the traditional concept of a regulated forest. Even for the cases when point equilibria were obtained, highly skewed age class distributions result, rather than the equal or nearly equal distributions that have historically characterized regulated forests.

Point equilibria, even with skewed age class distributions, usually lend themselves to equal annual or periodic treatments. Table 13.8 portrays an example of an area-regulated harvest schedule designed to maintain a skewed age class distribution in point equilibrium, as prescribed by the MAXJV objective function for the MHA = 2 model with deterministic rates of stand mortality included.[1] The harvest schedule maintains 22 percent of the forest at age 0–9 years, 17.6 percent at age 10–19 years, 33.04 percent at age 20–149 years, 8.93 percent at age 150–249 years, and 18.43 percent at age 250 years or older. It should be noted that final harvests are required on portions of the forest at ages considerably less than would typically be prescribed for optimal timber management. These harvests are used to maintain the skewed age class distribution prescribed for wildlife habitat instead. Although they are imposed in a point equilibrium fashion (by time period) and conform to the Clutter et al. definition of a regulated forest, these types of harvests would not normally be included in a traditional forest regulation schedule. Under this schedule, however, both habitat allocation and timber harvest (in an area control sense) are sustained over a long time frame.

Cyclic forms of equilibrium usually result in uneven (but recurrent) annual or periodic harvest levels. Such variations in periodic harvests and subsequent age class distributions do not conform to the Clutter et al. definition of a regulated forest. It is clear that traditional definitions of regulated forestry do not adequately address cyclic dynamic equilibrium cases, but it should be noted that this is not a problem for our traditional definitions of sustainability. A common definition of sustainabil-

---

1. These included regeneration failures, assumed for 20 percent of all new stands, and fire mortality rates ranging from 1.5 to 3.0 percent per decade, depending on stand age.

TABLE 13.8

Expected Harvest and Mortality Area by Age Class per Decade
Regulated over a 60-Decade Rotation

	Percentage of Forest		
Age in Decades	Harvested	Burned Outside of Harvests	Remaining
0	—	—	22.00
1	—	4.4000[a]	17.60
2	14.52	.0924	2.99
3	—	.0896	2.90
4	—	.0869	2.81
5	—	.0843	2.73
6	—	.0818	2.64
7	—	.0793	2.56
8	—	.0770	2.49
9	—	.0498	2.44
10	—	.0488	2.39
11	—	.0478	2.34
12	—	.0468	2.30
13	—	.0459	2.25
14	—	.0450	2.20
15	1.21	.0199	.98
16	—	.0195	.96
17	—	.0191	.94
18	—	.0188	.92
19	—	.0184	.90
20	—	.0180	.88
21	—	.0177	.86
22	—	.0173	.85
23	—	.0170	.83
24	—	.0166	.81
25	.13	.0137	.67
26	—	.0101	.66
27	—	.0099	.65
28	—	.0098	.64
29	—	.0096	.63
30	—	.0095	.62
.	.	.	.
.	.	.	.
.	.	.	.
58	—	.0062	.41
59	—	.0061	.40
60	.40	—	—

[a] This number represents regeneration failures for 20 percent of regenerated stands rather than fire mortality.

ity, from the Multiple Use—Sustained Yield Act of 1960 (16 U.S.C. 531), requires the continued maintenance of annual *or regular periodic* production of renewable resources. Such a definition easily accommodates all three deterministic forms of equilibria.

Appendix

Treating environmental site conditions such as soils and climate as constants, a broad, deterministic model of wildlife population dynamics in a forest area from a systems ecology perspective (see Kitching, 1983) might be stated as follows. Let j represent forest sites within a study area sufficiently large for determining wildlife habitat quality. Rates of change over time could be modeled by

$$\frac{dF_{ij}}{dt} = f_{ij}(F, X, n) \qquad \forall i, j \qquad F = [F_{ij}]_{I \times J} \qquad (13.13)$$

$$X = [X_{ij}]_{I \times J}$$

$$n = [n_{jk}]_{J \times K},$$

$$\frac{dn_{jk}}{dt} = g_k(n, C, W) \qquad \forall j, k \qquad C = [C_{jk}]_{J \times K} \qquad (13.14)$$

$$W = [W_{jk}]_{J \times K},$$

with

$$C_{jk} = q_k(H_{jk}) = h_k(\vec{F}_j) \qquad \forall j, k, \qquad (13.15)$$

$$N_k = \sum_j n_{jk} \qquad \forall k, \qquad (13.16)$$

where

F_{ij} = the ith forest cover attribute of importance to wildlife at forest site j,

f_{ij} = site specific functions of forest attribute change,

X_{ij} = a forest cover management variable that alters attribute i at site j,

n_{jk} = the population of wildlife species k at site j,

g_k = wildlife species population growth functions,

C_{jk} = the wildlife population carrying capacity for species k at site j,

W_{jk} = a wildlife population management variable that alters the population of species k at site j, and

$\vec{F}_j$ = the vector of all forest cover attributes at site j.

Note in (13.15) that instead of expressing carrying capacity as a function

q_k of habitat H_{jk}, a composite function h_k can be used to determine carrying capacity directly from forest cover attributes. In systems ecology terminology F_{ij}, n_{jk}, and C_{jk} are state variables endogenous to the dynamic system and X_{ij} and W_{jk} are exogenous driving (or control) variables.

Because the focus of this chapter is long-term habitat management, a number of simplifications are made. First, extreme management practices that might destabilize vegetative communities are not considered, so endemic wildlife species are assumed to have little effect on forest vegetation. Equation (13.13) becomes

$$\frac{dF_{ij}}{dt} = f_{ij}(\mathbf{F}, \mathbf{X}) \qquad \forall\, i,\, j. \tag{13.17}$$

Similarly, the influence of offsite vegetation on onsite forest cover is disregarded. Equation (13.17) becomes

$$\frac{dF_{ij}}{dt} = f_{ij}(\vec{F_j}, \vec{X_j}) \qquad \forall\, i,\, j, \tag{13.18}$$

where $\vec{X_j}$ is the vector of management variables for all forest cover attributes at site j.

Many attributes of forest cover change somewhat slowly compared to the faster responses of wildlife populations. Consequently, forest cover is often modeled using longer time steps. Given our emphasis on long-term habitat management over a large area, many of the local, short-term fluctuations in wildlife populations might be reduced to independent average long-term dynamics. On this basis, equation (13.14) simplifies to

$$\frac{dn_{jk}}{dt} = g_k(\vec{n_k}, \vec{C_k}) \qquad \forall\, j,\, k, \tag{13.19}$$

where $\vec{n_k}$ is the vector of species k population variables across all sites, and $\vec{C_k}$ is the vector of species k carrying capacities across all sites.

Equation (13.19) retains elements for modeling spatial population interactions and growth over time in order to simultaneously optimize habitat allocation and arrangement as in previous chapters. However, the complexity of this formulation is difficult to retain for the longer planning horizons required to examine sustainability. Consequently, we disregard site-specific population levels in favor of simply modeling overall population growth by species. This also precludes other site-specific population dynamics, such as local species interactions. We thus eliminate equation (13.16) and reduce equation (13.19) to

$$\frac{dN_k}{dt} = g_k(N_k, \sum_j C_{jk}) \qquad \forall\, k. \tag{13.20}$$

Discretizing time in equations (13.15), (13.18), and (13.20) into P time periods (Δt in length and indexed by t leads to the following system of equations:

$$F_{ij,t+\Delta t} = F_{ijt} + f_{ij}(\vec{F}_{jt}, \vec{X}_{jt})\Delta t \qquad \forall\, i,\, j \tag{13.21}$$
$$t = 1, \ldots, P - 1,$$

$$C_{jkt} = h_k(\vec{F}_{jt}) \qquad \forall\, j,\, k,\, t, \tag{13.22}$$

$$N_{k,t+\Delta t} = N_{kt} + g_k(N_{kt}, \sum_j C_{jkt})\Delta t \qquad \forall\, k, \tag{13.23}$$
$$t = 1, \ldots, P - 1.$$

Beginning from some set of initial conditions, equations (13.21)–(13.23) could be used as the basis of a simulation model. By combining these equations with an appropriate objective function and terminal constraints (or a long planning horizon), these same equations are the foundation of a discrete-time, optimal control nonlinear programming model (Luenberger, 1984) and the linear model described in chapter 13. In this context, the variables for managing forest cover, X_{ijt}, would be the control variables, or the choice variables in our model.

Chapter 14

SYNTHESIS

Our last task is to step back and synthesize the variety of material on spatial optimization from the preceding four parts. In part I, we developed approaches to optimizing systems for which the spatial relationships are directly known, a priori. This might result from observing equilibrium metapopulation numbers for a species in habitat complexes that exhibit a range of connectivity between patches (see Hanski, 1994) or from observing other such patterns including human behavior (as in chapter 4). In chapters 2 and 3 we also presented the discrete cellular and continuous geometric spatial definitions that were then used throughout most of the book. In part II (chapter 5), we extended the methods from part I to address the more general case in which the spatial relationships are still known, a priori, but they are not known with certainty. In this context, we might or might not fully recognize the underlying reasons for the spatial relationships that we observe; when we observe them, it is as spatial autocorrelations between random variables. The wildlife model in chapter 6 further demonstrated how the methods from parts I and II can be usefully combined. The postoptimization procedures described in chapter 7 could be applied to any of the models in the book, except for those that already explicitly account for random technical coefficients. In part III we shifted from static to dynamic optimization models to address problems in which the spatial relationships are not known or estimated directly, but can be expected to result from the interaction of spatially defined ecosystem components through dynamic processes. The black-footed ferret case study presented in chapter 9 is a real-world application of the methods developed in chapter 8. The forest pest management model in chapter 10 uses a similar characterization of the problem, but is, in the final analysis, quite different because of the distinction between maximizing wildlife populations and minimizing pest populations. Likewise, the pest and stormflow models in chapters 10 and 11 require integer solvers and are not intrinsically integer-friendly, making real-world studies intractable without heuristic solution methods. Part IV introduced more complicated objective functions for optimizing sustainable species richness to highlight the importance of examining new kinds of objectives for managed ecosystems. By combining species richness objectives with the wildlife model from chapter 8, chapter 12 demonstrates how important such objectives can be in spatial optimization. But from the case study involving 92 terrestrial wildlife species in chapter 13, it is clear that the combination of many species, many time periods, piecewise approximation of nonlinear viability functions, and spatial detail creates a model too large for our current linear programming solvers. Beyond that, all our models require strong assumptions regarding ecosystem function and behavior, ignoring complications such as predator–prey relationships, population-dependent dispersal behavior, complex pest–host relationships, and nonlinear water flow dynamics.

Clearly some of the models we have presented are more ready for real-world application than others. Also, some of our methods can be combined, and some cannot. It was never our intention to present a single model or approach in this book. Rather, we have tried to pull together at least partial solutions to a variety of problems with the common thread of a significant spatial component. Individual planning models typically highlight particular problems, and no single model will ever address all possible problems.

An Adaptive Management Context

Perhaps the most important topic to discuss in this synthesis is the context for methods such as those presented here. Whereas computing technology limitations are diminishing at an amazing rate, we will probably never know as much as we would want about the ecosystems we must manage. A natural reaction to the material presented in this book would be that we simply do not know enough about ecological systems to actually optimize them. Our reply is that our methods must be applied in the context of an adaptive learning process (and we have noted this context several times). We would never argue that our knowledge level is adequate to find a permanent optimal strategy for managing an ecosystem with a one-time optimization analysis. At the same time, we would argue that an adaptive management process that does not take advantage of optimization methods is much less likely to make progress either in learning about the ecological system or in managing it.

To demonstrate, suppose we have 100 land units (for example, in a 10×10 grid) to manage, and the spatial configuration of our management actions is important. Even if we must consider only one action (versus none), with no scheduling component, there are still 2^{100} or 1.2676×10^{30} possible spatial layouts. Even if 99.9999999 percent (all but a trillionth) of the layouts can be eliminated as undesirable, we still have 1.2676×10^{21} options. Even if there are a trillion layouts that are acceptable, we have only a 7.8886×10^{-13} $(1 \times 10^{9} \div 1.2676 \times 10^{21})$ chance of hitting an acceptable solution on our first try if we randomly arrange our management actions. With such a tiny probability, a strict trial-and-error approach using either simulation or the actual ecosystem as a guinea pig is not likely to be successful. Naturally, if we learn a great deal from each trial, we might be able to beat these odds eventually, but there are at least three factors that limit the rate at which we might converge on an acceptable solution. First, with managed ecosystems, the time frame for the feedback cycle may be very long. Adaptive management was originally conceptualized in the context of marine fishing, where an annual cycle provides feedback on the population response to harvest levels (Holling, 1978; Walters, 1986). In other ecosystems, however, it may take many years before the consequences of management actions manifest themselves, and even annual feedback from a stochastic process could take a long time to interpret correctly. Second, the problem of obtaining reliable feedback is not trivial. If we begin from a position of minimal understanding, it is likely that we are also unsure about how to monitor ecosystem response. We might measure the wrong thing or mismeasure the right thing for the same reasons that we are unsure of the ecosystem response in the first place. Third, many systems may exhibit behavior that resists convergence based on a trial-and-error approach. For example, figure 14.1 depicts the classic cobweb model. Assuming that we wish to maximize net benefits by equating marginal costs (MC) with marginal benefits (MB), we attempt to adjust the control variable, C, by trial and error. Also, assume that we have a good understanding of the MC function, but must observe MB from the system after setting C. If we start at C^{0}, we observe that $MB > MC$, so we increase C to C', where MC equals our observed MB at C^{0}. But now, we observe that $MC > MB$, so we decrease C to C'', where MC equals our observed MB at C'. Now $MB > MC$ again, so we increase C to C''',

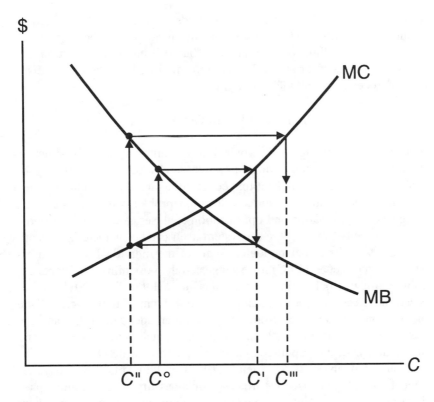

FIGURE 14.1
An unstable cobweb model.

where MC equals MB at C''. Obviously this process (even though deterministic) continues to explode with worse and worse results relative to the desired solution, where MC = MB (see Chiang, 1984, for further details). Many other equilibrium systems can show similar instabilities, including ones with multiple interactive equilibria. Systems with bifurcations, such as the catastrophe and chaotic systems, may also confound trial-and-error processes.

The upshot is that adaptive management is a powerful concept, but it should not be viewed as a replacement for analysis. Up-front analysis is inadequate without follow-up monitoring, reanalysis, and management plan revision. At the same time, using an adaptive process is no excuse for ceasing attempts at understanding the system as well as we possibly can or for making mistakes that could have been prevented with more careful analysis. We visualize the methods in this book being used in combination with other analyses, and in an adaptive process that acknowledges our limited understanding but still emphasizes optimizing the system at each turn with the best information, formulations, and solution procedures available at that time.

Simulation Versus Optimization

For our closing thought, we would like to comment on the distinction between simulation and optimization. Obviously, our optimization models contain a large component that could be regarded as simulation. Furthermore, most natural resource allocation problems have aspects that suggest both simulation and optimization analyses. Also, it is quite possible that the behavior of the ecosystem itself can be characterized as optimizing. At the smaller biological scale of the individual, investigators such as Bloom et al. (1985) have suggested the possibility that plants exhibit optimizing behavior. Along similar lines, Hof et al. (1990) hypothesized that trees attempt to maximize the minimum of two functions: the net carbon gain achieved from leaf area given adequate water, and the net carbon gain from root biomass given adequate carbon dioxide and light energy. At the population or metapopulation scale, discretized biodiffusion models such as Allen's (1987) and the black-footed ferret model presented in chapter 9 are based on an assumption of limited maximizing behavior by animal populations. At the community or ecosystem scale, the process of natural selection is certainly an optimization process in the sense that it eliminates inefficient components in the system. It also seems reasonable that the ecosystem components that remain do so because they function efficiently. Given enough time, the system that survives might then be expected to exhibit optimizing behavior as well. At that point, the distinction between simulation and optimization could be quite subtle. If ecosystems optimize themselves, then optimization might be used to simulate ecosystem behavior. Likewise, if we attempt to optimize the variables that humans control, we must include a simulation of the ecosystem's response. By providing testable hypotheses, optimization models could be used along with simulation models to direct behavioral research in the biological sciences (Hof et al., 1990).

Different disciplines have tended to focus on certain types of analysis, but our conclusion is that resource economists and management scientists probably need to pay more attention to simulation methodologies and ecologists probably need to pay more attention to optimization methodologies. As Rapport and Turner (1977) wrote over 20 years ago,

Ecological processes have traditionally been studied from several vantage points. . . . None of these approaches, however, explicitly address what some . . . have regarded as one of the central problems of ecology—the ways in which scarce resources are allocated among alternative uses and users. The question is, of course, fundamental to economic thinking . . . and it is for this reason that we have recently seen the introduction of essentially economic models and modes of thought in ecology.

Progress over the last 2 decades in this direction has been occurring slowly. We hope this book provides some small impetus for greater scientific movements along these lines.

REFERENCES

Abadie, J. 1978. The GRG method of nonlinear programming. In H. J. Greenberg, ed., *Design and Implementation of Optimization Software*, pp. 335–363. Amsterdam: Sijthoff and Noordhoff.

Abramowitz, M., and I. A. Stegun. 1965. *Handbook of Mathematical Functions with Formulas, Graphs, and Mathematical Tables*. Washington, D.C.: National Bureau of Standards.

Agee, J. K. 1991. Fire history of Douglas-fir forests in the Pacific Northwest. In L. F. Ruggiero, K. B. Aubrey, A. B. Carey, and M. H. Huff, tech. coords., *Wildlife and Vegetation of Unmanaged Douglas-Fir Forests*. General Technical Report PNW-GTR-285, pp. 25–33. Portland, Ore.: USDA Forest Service, Pacific Northwest Research Station.

Allen, L. J. S. 1983. Persistence and extinction in single-species reaction-diffusion models. *Bulletin of Mathematical Biology* 45: 209 227.

Allen, L. J. S. 1987. Persistence, extinction, and critical patch number for island populations. *Journal of Mathematical Biology* 24: 617–625.

Allen, T. G. H., and T. W. Hoekstra. 1992. *Toward a Unified Ecology*. New York: Columbia University Press.

Amman, G. D. 1988. *Proceeding: Symposium on the Management of Lodgepole Pine to Minimize Losses to the Mountain Pine Beetle*. Intermountain Research Station General Technical Report INT-262. Ogden, Utah: USDA Forest Service.

Anderson, E., S. C. Forrest, T. W. Clark, and L. Richardson. 1986. Paleobiology, biogeography, and systematics of the black-footed ferret, *Mustela nigripes* (Audubon and Bachman) 1851. *Great Basin Naturalist Memoirs* 8: 11–62.

Apa, A. D., D. W. Uresk, and R. L. Linder. 1990. Black-tailed prairie dog populations one year after treatment with rodenticides. *Great Basin Naturalist* 50: 107–113.

Austin, R. F., 1984. Measuring and comparing two-dimensional shapes. In G. L. Gaile and C. J. Willmontt, eds., *Spatial Statistics and Models*, pp. 293–312. Boston: D. Reidel.

Balintfy, J. L. 1970. Nonlinear programming for models with joint chance constraints. In J. Abadie, ed., *Integer and Nonlinear Programming,* pp. 337–352. Amsterdam: North-Holland.

Bennett, F. A. 1970. *Yields and Stand Structural Patterns for Old-Field Plantations of Slash Pine.* USDA Forest Service Research Paper SE-60. Asheville, N.C.: Southeast Forest Experiment Station.

Bevers, M., J. Hof, B. Kent, and M. G. Raphael. 1995. Sustainable forest management for optimizing multispecies wildlife habitat: a coastal Douglas-fir example. *Natural Resource Modeling* 9: 1–23.

Bevers, M., J. Hof, and C. Troendle. 1996. Spatially optimizing forest management schedules to meet stormflow constraints. *Water Resources Bulletin* 32(5): 1007–1015.

Bevers, M., J. Hof, D. W. Uresk, and G. L. Schenbeck. 1997. Spatial optimization of prairie dog colonies for black-footed ferret recovery. *Operations Research* 45(4): 495–507.

Bevers, M., and B. M. Kent. 1991. Generalized harvest scheduling in FORPLAN Version 2: an update. In M. A. Buford, compiler, *Proceedings of the 1991 Symposium on Systems Analysis in Forest Resources.* General Technical Report SE-GTR-74, pp. 297–299. Asheville, N.C.: USDA Forest Service, Southeastern Forest Experiment Station.

Biggins, D. E., B. J. Miller, L. R. Hanebury, B. Oakleaf, A. H. Farmer, R. Crete, and A. Dood. 1993. A technique for evaluating black-footed ferret habitat. *U.S. Fish and Wildlife Service Biological Report* 93(13): 73–88.

Biggins, D. E., M. H. Schroeder, S. C. Forrest, and L. Richardson. 1986. Activity of radio-tagged black-footed ferrets. *Great Basin Naturalist Memoirs* 8: 135–140.

Bloom, A. J., F. S. Chapin, and H. A. Mooney. 1985. Resource limitation in plants: an economic analogy. *Annual Review of Ecology and Systematics* 16: 363–392.

Bowers, M. A., and L. C. Harris. 1994. A large-scale metapopulation model of interspecific competition and environmental change. *Ecological Modelling* 72: 251–273.

Bowes, M. D., and J. B. Loomis. 1980. A note on the use of travel-cost models with unequal zonal populations. *Land Economics* 56: 465–470.

Brown, G. W., and J. T. Krygier. 1970. Effects of clearcutting on stream temperature. *Water Resources Research* 6: 1133–1140.

Brown, G. G., and H. C. Rutemiller. 1977. Means and variances of stochastic vector products with applications to random linear models. *Management Science* 24(2): 210–216.

Burkey, T. V. 1989. Extinction in nature reserves: the effect of fragmentation and the importance of migration between reserve fragments. *Oikos* 55: 75–81.

Burt, O. R., and D. Brewer. 1971. Estimation of net social benefits from outdoor recreation. *Econometrica* 39: 813–827.

Cesario, F. J. 1976. Value of time in recreation benefit studies. *Land Economics* 52: 32–41.

Charnes, A., and W. W. Cooper. 1963. Deterministic equivalents for optimizing and satisficing under chance constraints. *Operations Research* 11: 18–39.

Chiang, A. C. 1984. *Fundamental Methods of Mathematical Economics.* New York: McGraw-Hill.

Cicchetti, C. J., A. C. Fisher, and V. K. Smith. 1976. An econometric evaluation of a generalized consumer surplus measure: the mineral king controversy. *Econometrica* 44: 1259–1276.

Cincotta, R. P. 1985. Habitat and dispersal of black-tailed prairie dogs in the Badlands National Park. Unpublished M.S. thesis, Colorado State University, Fort Collins.

Cincotta, R. P., D. W. Uresk, and R. M. Hansen. 1987. Demography of black-tailed prairie dog populations reoccupying sites treated with rodenticide. *Great Basin Naturalist* 47: 339–343.

Cincotta, R. P., D. W. Uresk, and R. M. Hansen. 1988. A statistical model of expansion in a colony of black-tailed prairie dogs. In D. W. Uresk, G. L. Schenbeck, and R. Cefkin, tech. coords., *Eighth Great Plains Wildlife Damage Control Workshop Proceedings, April 28–30, 1987.* General Technical Report RM-154, pp. 30–33. Rapid City, S.D.: Rocky Mountain Forest and Range Experiment Station.

Clark, T. W. 1989. Conservation biology of the black-footed ferret, *Mustela nigripes.* Special Scientific Report No. 3. Philadelphia, Pa.: Wildlife Preservation Trust International.

Clawson, M., and J. Knetsch. 1966. *Economics of Outdoor Recreation.* Baltimore: John Hopkins University Press.

Clutter, J., J. Fortson, L. Pienar, G. Brister, and R. Bailey. 1983. *Timber Management: A Quantitative Approach.* New York: Wiley.

Cole, W. E., and M. D. McGregor. 1983. *Estimating the Rate and Amount of Tree Loss from Mountain Pine Beetle Infestations.* Intermountain Research Station Research Paper INT-318. Ogden, Utah: USDA Forest Service.

Czaplewski, R. L., R. M. Reich, and W. A. Bechtold. 1994. Spatial autocorrelation in growth of undisturbed natural pine stands across Georgia. *Forest Science* 40: 314–328.

den Boer, P. J. 1981. On the survival of populations in a heterogeneous and variable environment. *Oecologia* 50: 39–53.

Diamond, J. M. 1975. The island dilemma: lessons of modern biogeographic studies for the design of natural reserves. *Biological Conservation* 7: 129–146.

Diamond, J. M. 1976. Island biogeography and conservation: strategy and limitations. *Science* 193: 1027–1029.

Duloy, J. H., and R. D. Norton. 1975. Prices and incomes in linear programming models. *American Journal of Agricultural Economics* 57: 591–600.

Dunne, T., and L. B. Leopold. 1978. *Water in Environmental Planning.* San Francisco: Freeman.

Dykstra, D. P. 1984. *Mathematical Programming for Natural Resource Management.* New York: McGraw-Hill.

Eyre, F. H., ed. 1980. *Forest Cover Types of the United States and Canada.* Washington, D.C.: Society of American Foresters.

Fahrig, L., and G. Merriam. 1985. Habitat patch connectivity and population survival. *Ecology* 66: 1762–1768.

Fahrig, L., and G. Merriam. 1994. Conservation of fragmented populations. *Conservation Biology* 8: 50–59.

Fahrig, L., and J. Paloheimo. 1988. Effect of spatial arrangement of habitat patches on local population size. *Ecology* 69: 468–475.

FEMAT. 1993. *Forest Ecosystem Management: An Ecological, Economic, and Social Assessment.* Report of the Forest Ecosystem Management Assessment Team. Washington, D.C.: U.S. Government Printing Office publication 1993–793–071.

Ford-Robertson, F. C., ed. 1971. *Terminology of Forest Science, Technology, Practice and Products.* Washington, D.C.: Society of American Foresters.

Forrest, S. C., T. W. Clark, L. Richardson, and T. M. Campbell, III. 1985. *Black-Footed Ferret Habitat: Some Management and Reintroduction Considerations.* Wyoming Bureau of Land Management Technical Bulletin No. 2.

Fox, K. A., J. K. Sengupta, and E. Thorbecke. 1966. *The Theory of Quantitative Economic Policy with Applications to Economic Growth and Stabilization.* Amsterdam: North-Holland.

Franklin, J. F., and R. T. Forman. 1987. Creating landscape patterns by forest cutting: ecological consequences and principles. *Landscape Ecology* 1: 5–18.

Game, M. 1980. Best shape for nature reserves. *Nature* 287: 630–632.

Gill, P. E., W. Murray, and M. H. Wright. 1981. *Practical Optimization.* San Diego: Academic Press.

Gilpin, M. E. 1987. Spatial structure and population vulnerability. In M. E. Soulé, ed., *Viable Populations for Conservation,* pp. 125–139. Cambridge, England: Cambridge University Press.

Goodman, D. 1987. The demography of chance extinction. In M. E. Soulé, ed., *Viable Populations for Conservation,* pp. 11–34. Cambridge, England: Cambridge University Press.

Gurney, W. S. C., and R. M. Nisbet. 1975. The regulation of inhomogeneous populations. *Journal of Theoretical Biology* 52: 441–457.

Hamel, D. R., and C. I. Shade. 1987. *Pesticide Use in Forest Management.* USDA Forest Service publication FS-404, Forest Pest Management. Washington, D.C.

Hanski, I. 1991. Single-species metapopulation dynamics: concepts, models and observations. *Biological Journal of the Linnean Society* 42: 17–38.

Hanski, I. 1994. A practical model of metapopulation dynamics. *Journal of Animal Ecology* 63: 151–162.

Harris, L. D. 1984. *The Fragmented Forest.* Chicago: The University of Chicago Press.

Harris, L. D., and P. B. Gallagher. 1989. New initiatives for wildlife conservation: the need for movement corridors. In G. Mackintosh, ed., *Preserving Communities and Corridors,* pp. 11–34. Washington, D.C.: Defenders of Wildlife.

Harris, R. B., T. W. Clark, and M. L. Shaffer. 1989. Extinction probabilities for isolated black-footed ferret populations. In U. S. Seal, E. T. Thorne, M. A. Bogan, and S. H. Anderson, eds., *Conservation Biology and the Black-Footed Ferret,* pp. 69–82. New Haven, Conn.: Yale University Press.

Harvey, C. M. 1985. Decision analysis models for social attitudes toward inequity. *Management Science* 31(10): 1199–1212.

Heal, G. M. 1973. *The Theory of Economic Planning.* Amsterdam: North Holland.

Henderson, F. R., P. F. Springer, and R. Adrian. 1969. *The Black-Footed Ferret in South Dakota.* Technical Bulletin No. 4. Pierre: South Dakota Department of Game, Fish and Parks.

Hermann, R. K., and D. P. Lavender. 1990. *Pseudotsuga menziesii.* In R. M. Burns and B. H. Honkala, tech. coords., *Silvics of North America,* Volume 1: *Conifers,* pp. 527–540. Agriculture Handbook 654. Washington, D.C.: USDA Forest Service.

Hewlett, J. D., and A. R. Hibbert. 1967. Factors affecting the response of small watersheds to precipitation in humid regions. In W. E. Sopper and H. W. Lull, eds., *Forest Hydrology,* pp. 275–290. Oxford, England: Pergamon Press.

Hillman, C. N., R. L. Linder, and R. B. Dahlgren. 1979. Prairie dog distributions in areas inhabited by black-footed ferrets. *American Midland Naturalist* 102: 185–187.

Hof, J. 1993. *Coactive Forest Management.* San Diego: Academic Press.

Hof, J., M. Bevers, L. Joyce, and B. Kent. 1994. An integer programming approach for spatially and temporally optimizing wildlife populations. *Forest Science* 40(1): 177–191.

Hof, J., M. Bevers, and B. Kent. 1997. An optimization approach to area-based forest pest management over time and space. *Forest Science* 43(1): 121–128.

Hof, J., M. Bevers, and J. Pickens. 1995. Pragmatic approaches to optimization with random yield coefficients. *Forest Science* 41(3): 501–512.

Hof, J., M. Bevers, and J. Pickens. 1996. Chance-constrained optimization with spatially autocorrelated forest yields. *Forest Science* 42(1): 118–123.

Hof, J., and C. H. Flather. 1996. Accounting for connectivity and spatial correlation in the optimal placement of wildlife habitat. *Ecological Modelling* 88: 143–155.

Hof, J. G., and L. A. Joyce. 1992. Spatial optimization for wildlife and timber in managed forest ecosystems. *Forest Science* 38(3): 489–508.

Hof, J. G., and L. A. Joyce. 1993. A mixed integer linear programming approach for spatially optimizing wildlife and timber in managed forest ecosystems. *Forest Science* 39(4): 816–834.

Hof, J. G., B. M. Kent, and J. B. Pickens. 1992. Chance constraints and chance maximization with random yield coefficients in renewable resource optimization. *Forest Science* 38(2): 305–323.

Hof, J. G., and D. A. King. 1982. On the necessity of simultaneous recreation demand equation estimation. *Land Economics* 58: 547–552.

Hof, J. G., and J. B. Loomis. 1983. A recreation optimization model based on the travel cost method. *Western Journal of Agricultural Economics* 8(1): 76–85.

Hof, J. G., and J. B. Pickens. 1991. Chance-constrained and chance-maximizing mathematical programs in renewable resource management. *Forest Science* 37(1): 308–325.

Hof, J. G., and M. G. Raphael. 1993. Some mathematical programming approaches for optimizing timber age class distributions to meet multispecies wildlife population objectives. *Canadian Journal of Forest Research* 23: 828–834.

Hof, J., D. Rideout, and D. Binkley. 1990. Carbon fixation in trees as a micro optimization process: an example of combining ecology and economics. *Ecological Economics* 2: 243–256.

Hof, J. G., K. S. Robinson, and D. R. Betters. 1988. Optimization with expected values of random yield coefficients in renewable resource linear programs. *Forest Science* 34(3): 634–646.

Holling, C. S. 1978. *Adaptive Environmental Assessment and Management.* Chichester, England: Wiley.

Hoogland, J. L., D. K. Angell, J. G. Daley, and M. C. Radcliffe. 1988. Demography and population dynamics of prairie dogs. In D. W. Uresk, G. L. Schenbeck, and R. Cefkin, tech. coords., *Eighth Great Plains Wildlife Damage Control Workshop Proceedings, April 28–30, 1987.* General Technical Report RM-154, pp. 18–22. Rapid City, S.D.: Rocky Mountain Forest and Range Experiment Station.

Horton, R. E., 1935. *Surface Runoff Phenomena. Part I: Analysis of the Hydrograph.* Horton Hydrological Laboratory Publication 101. New York: Voorheesville.

Houston, B. R., T. W. Clark, and S. C. Minta. 1986. Habitat suitability index model for the black-footed ferret: a method to locate transplant sites. *Great Basin Naturalist Memoirs* 8: 99–114.

Hursh, C. R., and E. F. Brater. 1944. Separating hydrographs in surface- and subsurface-flow. *Transactions of the American Geophysical Union* 22: 863–871.

Jagannathan, R. 1974. Chance-constrained programming with joint constraints. *Operations Research* 22: 358–372.

Janda, R. J., M. K. Nolan, D. R. Harden, and S. M. Colman. 1975. *Watershed Conditions in the Drainage Basin of Redwood Creek, Humboldt Co., California, as of 1973.* Open File Report no. 75.568. Menlo Park, Calif.: U.S. Geological Survey.

Johnson, N. L., and S. Kotz. 1970. Continuous Univariate Distributions: 2. New York: Wiley.

Johnson, K. N., and H. L. Scheurman. 1977. Techniques for prescribing optimal timber harvest and investment under different objectives: discussion and synthesis. *Forest Science Monograph* 18.

Julien, P. Y., B. Saghafian, and F. L. Ogden. 1995. Raster-based hydrologic modeling of spatially-varied surface runoff. *Water Resources Bulletin* 31(3): 523–536.

Just, R. E., D. Hueth, and A Schmitz. 1982. *Applied Welfare Economics and Public Policy.* Englewood Cliffs, N.J.: Prentice Hall.

Kareiva, P. 1990. Population dynamics in spatially complex environments: theory and data. *Philosophical Transactions of the Royal Society of London B* 330: 175–190.

Kaufman, W. 1993. How nature really works. *American Forests* 99(3–4): 17–19, 59–61.

Keeney, R. L. 1974. Multiplicative utility functions. *Operations Research* 21: 22–34.

Kierstead, H., and L. B. Slobodkin. 1953. The size of water masses containing plankton blooms. *Journal of Marine Research* 12: 141–147.

Kitching, R. L. 1983. *Systems Ecology, An Introduction to Ecological Modelling.* St. Lucia: University of Queensland Press.

Knowles, C. J. 1985. Population recovery of black-tailed prairie dogs following control with zinc phosphide. *Journal of Range Management* 39: 249–251.

Liebhold, A. M., W. L. MacDonald, D. Bergdahl, and V. C. Mastro. 1995. Invasion by exotic forest pests: a threat to forest ecosystems. *Forest Science Monograph* 30.

Linder, R. L., R. B. Dahlgren, and C. N. Hillman. 1972. Black-footed ferret–prairie dog interrelationships. In *Symposium on Rare and Endangered Wildlife of the Southwestern United States, September 22–23, Albuquerque, N.M.,* pp. 22–37. Santa Fe: New Mexico Department of Game and Fish.

Long, G. E. 1977. Spatial dispersion in a biological control model for larch casebearer. *Environmental Entomology* 6(6): 843–852.

Luenberger, D. G. 1984. *Introduction to Linear and Nonlinear Programming.* Reading, Mass.: Addison-Wesley.

Lull, H. W., and K. G. Reinhart. 1972. *Forest and Floods in the Eastern United States.* Radnor, Pa.: USDA Forest Service Research Paper NE-226.

Mace, G. M., and R. Lande. 1991. Assessing extinction threats: toward a reevaluation of IUCN threatened species categories. *Conservation Biology* 5: 148–157.

Marcot, B. G. 1984. Habitat relationships of birds and young-growth Douglas fir in northwestern California. Ph.D. dissertation, Oregon State University, Corvallis.

Marcot, B. G., R. Holthausen, and H. Salwasser. 1986. Viable population planning. In B. A. Wilcox, P. F. Brussard, and B. G. Marcot, eds., *The Management of Viable Populations: Theory, Applications and Case Studies,* pp. 49–61. Stanford, Calif.: Center for Conservation Biology, Stanford University.

Margules, C. R., G. A. Milkovits, and G. T. Smith. 1994. Contrasting effects of habitat fragmentation on the scorpion *Cercophonius squama* and an amphipod. *Ecology* 75: 2033–2042.

Markowitz, H. 1959. *Portfolio Selection: Efficient Diversification of Investment.* New York: Wiley.

Martin, L. J. 1981. Quadratic single and multi-commodity models of spatial equilibrium: A simplified exposition. *Canadian Journal of Agricultural Economics* 29: 21–48.

Matérn, B. 1986. *Spatial Variation.* New York: Springer-Verlag.

Matérn, B. 1993. On spatial statistics in forestry. In M. Köhl and G. Z. Gertner, eds., *Statistical Methods, Mathematics, and Computers,* pp. 18–26. Birmensdorf: Swiss Federal Institute for Forest, Snow and Landscape Research.

May, R. M. 1973. *Stability and Complexity in Model Ecosystems.* Princeton, N.J.: Princeton University Press.

McCarl, B. A., and T. H. Spreen. 1980. Price endogenous mathematical programming as a tool for sector analysis. *American Journal of Agricultural Economics* 62: 87–102.

McDonald, P. M. 1995. *Black-footed ferret monitoring, Buffalo Gap National Grassland, Winter 1994.* Unpublished report. Chadron: Nebraska National Forest.

McNeely, J. A., K. R. Miller, W. V. Reid, R. A. Mittermeir, and T. B. Werner. 1990. Strategies for conserving biodiversity. *Environment* 32(3): 16–40.

Merriam, G., K. Henein, and K. Stuart-Smith. 1991. Landscape dynamics models. In M. G. Turner and R. H. Gardner, eds., *Quantitative Methods in Landscape Ecology,* pp. 399–416. New York: Springer-Verlag.

Miller, B. J., G. E. Menkens, and S. H. Anderson. 1988. A field habitat model for black-footed ferrets. In D. W. Uresk, G. L. Schenbeck, and R. Cefkin, tech. co-ords., *Eighth Great Plains Wildlife Damage Control Workshop Proceedings, April 28–30, 1987,* General Technical Report RM-154, pp. 98–102. Rapid City, S.D.: Rocky Mountain Forest and Range Experiment Station.

Miller, B. L., and H. M. Wagner. 1965. Chance constrained programming with joint constraints. *Operations Research* 13: 930–945.

Milne, B. T., K. Johnston, and R. T. T. Forman. 1989. Scale-dependent proximity of wildlife habitat in a spatially-neutral Bayesian model. *Landscape Ecology* 2: 101–110.

Minta, S., and T. W. Clark. 1989. Habitat suitability analysis of potential transloca-tion sites for black-footed ferrets in north-central Montana. In T. W. Clark, D. Hinckley, and T. Rich, eds., *The Prairie Dog Ecosystem: Managing for Biologi-cal Diversity,* pp. 29–46. Billings: Montana Bureau of Land Management.

Musgrave, G. W., and H. N. Holtan. 1964. Infiltration. In Ven te Chow, ed., *Hand-book of Applied Hydrology,* Section 12. New York: McGraw-Hill.

Oakleaf, B., B. Luce, and E. T. Thorne. 1992. Evaluation of black-footed ferret rein-troduction in Shirley Basin, Wyoming. In B. Oakleaf, B. Luce, E. T. Thorne, and S. Torbit, eds., *1991 Annual Completion Report,* pp. 196–240. Cheyenne: Wyoming Game and Fish Department.

Oakleaf, B., B. Luce, and E. T. Thorne. 1993. An evaluation of black-footed ferret reintroduction in Shirley Basin, Wyoming. In B. Oakleaf, B. Luce, E. T. Thorne, and B. Williams, eds., *1992 Annual Completion Report,* pp. 220–234. Cheyenne: Wyoming Game and Fish Department.

Okubo, A. 1980. *Diffusion and Ecological Problems: Mathematical Models. Bio-mathematics 10.* New York: Springer-Verlag.

Orians, G. H. 1991. Preface. *American Naturalist* 137(supplement): S1–S4.

Paine, T. D., F. M. Stephen, and H. A. Taha. 1984. Conceptual model of infestation probability based on bark beetle abundance and host tree susceptibility. *Environ-mental Entomology* 13(3): 619–624.

Pickens, J. B., and P. E. Dress. 1988. Use of stochastic production coefficients in linear programming models: objective function distribution, feasibility, and dual activities. *Forest Science* 34: 574–591.

Pielou, E. C. 1977. *Mathematical Ecology.* New York: Wiley.

Plane, D. R., and C. McMillan, Jr. 1971. *Discrete Optimization: Integer Program-ming and Network analysis for Management Decisions.* Englewood Cliffs, N.J.: Prentice Hall.

Pluhowski, E. J., 1970. *Urbanization and Its Effects on the Temperature of the Streams on Long Island, N.Y.* U.S. Geological Survey Professional Paper 627-D.

Plumb, G. E., P. M. McDonald, and D. Searls. 1994. *Black-footed ferret reintroduc-tion in South Dakota: project description and 1994 protocol.* Unpublished report, Badlands National Park.

Polymenopoulos, A. D., and G. Long. 1990. Estimation and evaluation methods for population growth models with spatial diffusion: dynamics of mountain pine bee-tle. *Ecological Modelling* 51: 97–121.

Quinn, J. F., and A. Hastings. 1987. Extinction in subdivided habitats. *Conservation Biology* 1: 198–208.

Raphael, M. G. 1988. Long-term trends in abundance of amphibians, reptiles, and mammals in Douglas-fir forests of northwestern California. In R. C. Szaro, K. E. Severson, and D. R. Patton, tech. coords., *Management of Amphibians, Reptiles, and Small Mammals in North America*, General Technical Report RM-GTR-166, pp. 23–31. Fort Collins, Colo.: USDA Forest Service, Rocky Mountain Forest and Range Experiment Station.

Raphael, M. G., K. V. Rosenberg, and B. G. Marcot. 1988. Large-scale changes in bird populations of Douglas-fir forests, northwestern California. In J. A. Jackson, ed., *Bird Conservation 3*, pp. 63–83. Madison: University of Wisconsin Press.

Rapport, D. J., and J. E. Turner. 1977. Economic models in ecology. *Science* 195: 367–373.

Rawls, J. 1971. *A Theory of Justice.* Cambridge, Mass.: Belknap Press.

Reed, D. D., and H. E. Burkhart. 1985. Spatial autocorrelation of individual tree characteristics in loblolly pine stands. *Forest Science* 31: 575–587.

Reed, W. J., and D. Errico. 1987. Techniques for assessing the effects of pest hazards on long-run timber supply. *Canadian Journal of Forest Research* 17: 1455–1465.

Reeves, C. R., ed. 1993. *Modern Heuristic Techniques for Combinatorial Problems.* New York: Halsted Press.

Reid, W. V., and K. R. Miller. 1989. *Keeping Options Alive: The Scientific Basis for Conserving Biodiversity.* Washington, D.C.: World Resources Institute.

Richardson, L., T. W. Clark, S. C. Forrest, and T. M. Campbell, III. 1987. Winter ecology of black-footed ferrets (*Mustela nigripes*) at Meeteetse, Wyoming. *American Midland Naturalist* 117: 225–239.

Robichaud, P. R., and T. A. Waldrop. 1994. A comparison of surface runoff and sediment yields from low- and high-severity site preparation burns. *Water Resources Bulletin* 30(1): 27–34.

Roemer, D. M., and S. C. Forrest. 1996. Prairie dog poisoning in the northern Great Plains: an analysis of programs and policies. *Environmental Management* 20: 349–359.

Roessel, B. W. P. 1950. Hydrologic problems concerning the runoff in headwater regions. *Transactions of the American Geophysical Union* 31(3): 431–442.

Saunders, D. A., R. J. Hobbs, and C. R. Margules. 1991. Biological consequences of ecosystem fragmentation: a review. *Conservation Biology* 5: 18–32.

Schenbeck, G. L., and R. J. Myhre. 1986. Aerial photography for assessment of black-tailed prairie dog management on the Buffalo Gap National Grassland, South Dakota. USDA Forest Service FPM-MAG Report No. 86–7.

Schmeidler, D. 1969. The nucleolus of a characteristic function game. *SIAM Journal of Applied Mathematics* 17: 1163–1170.

Seal, U. S. 1989. Introduction. In *Conservation Biology and the Black-Footed Ferret*, pp. xi–xvii. New Haven, Conn.: Yale University Press.

Sengupta, J. K. 1972. *Stochastic Programming.* Amsterdam: North-Holland.

Sessions, J. 1992. Solving for habitat connections as a Steiner network problem. *Forest Science* 38(1): 203–207.

Shaffer, M. L. 1981. Minimum population sizes for species conservation. *BioScience* 31: 131–134.

Simberloff, D. 1988. The contribution of population and community biology to conservation science. *Annual Review of Ecology and Systematics* 19: 473–511.

Simberloff, D. S., and L. G. Abele. 1976. Island biogeography theory and conservation practice. *Science* 191: 285–286.

Simberloff, D., J. A. Farr, J. Cox, and D. W. Mehlman. 1992. Movement corridors: conservation bargains or poor investments? *Conservation Biology* 6(4): 493–504.

Skellam, J. G. 1951. Random dispersal in theoretical populations. *Biometrika* 38: 196–218.

Solow, A., S. Polasky, and J. Broadus. 1993. On the measurement of biological diversity. *Journal of Environmental Economics and Management* 24: 60–68.

Soulé, M. E. 1987. *Viable Populations for Conservation.* Cambridge, England: Cambridge University Press.

Soulé, M. E. 1991. Land use planning and wildlife maintenance. *Journal of the American Planning Association* 57(3): 313–323.

Temple, S. A., and J. R. Cary. 1988. Modeling dynamics of habitat-interior bird populations in fragmented landscapes. *Conservation Biology* 2: 340–347.

Thomas, J. W., tech. ed. 1979. *Wildlife Habitats in Managed Forests, the Blue Mountains of Oregon and Washington.* Agriculture Handbook 553. Washington, D.C.: USDA Forest Service.

Thomas, J. W., and H. Salwasser. 1989. Bringing conservation biology into a position of influence in natural resource management. *Conservation Biology* 3(2): 123–127.

Thorne, E. T., and D. W. Belitsky. 1989. Captive propagation and the current status of free-ranging black-footed ferrets in Wyoming. In *Conservation Biology and the Black-Footed Ferret,* pp. 223–234. New Haven, Conn.: Yale University Press.

Tilman, D., R. M. May, C. L. Lehman, and M. A. Nowak. 1994. Habitat destruction and the extinction debt. *Nature* 371: 65–66.

Troendle, C. A. 1985. Variable source area models. In M. G. Anderson and T. P. Burt, eds. *Hydrological Forecasting,* Chapter 12, pp. 347–403. New York: Wiley.

Troendle, C. A., and R. M. King, 1987. The effect of partial and clearcutting on streamflow at Deadhorse Creek, Colorado. *Journal of Hydrology* 90: 145–157.

U.S. Fish and Wildlife Service. 1994. *Black-Footed Ferret Reintroduction, Conata Basin/Badlands, South Dakota.* Final Environmental Impact Statement.

Uresk, D. W., J. G. MacCracken, and A. J. Bjugstad. 1981. Prairie dog density and cattle grazing relationships. In *Fifth Great Plains Wildlife Damage Control Workshop Proceedings, October 13–15, 1981, Lincoln, Nebraska,* pp. 199–201.

Uresk, D. W., and G. L. Schenbeck. 1987. Effect of zinc phosphide rodenticide on prairie dog colony expansion as determined from aerial photography. *Prairie Naturalist* 19: 57–61.

Van de Panne, C., and W. Popp. 1963. Minimum-cost cattle feed under probabilistic protein constraints. *Management Science* 9: 405–430.

Wagner, H. M. 1975. *Principles of Operations Research.* Englewood Cliffs, N.J.: Prentice Hall.

Walsh, R. G., R. A. Gillman, and J. B. Loomis. 1981. *Wilderness Resource Economics: Recreation Use and Preservation Values.* Colorado State University Department of Economics mimeo.

Walters, C. 1986. *Adaptive Management of Renewable Resources.* New York: Macmillan.

Weddell, B. J. 1991. Distribution and movements of Columbian ground squirrels (*Spermophilus columbianus* (Ord)): are habitat patches like islands? *Journal of Biogeography* 18: 385–394.

Weintraub, A., G. Jones, A. Magendzo, M. Meacham, and M. Kirby. 1994. A heuristic system to solve mixed integer forest planning models. *Operations Research* 42(6): 1010–1024.

Weintraub, A., and J. Vera. 1991. A cutting plane approach for chance constrained linear programs. *Operations Research* 39: 776–785.

Weitzman, M. L. 1992. On diversity. *Quarterly Journal of Economics* 107: 363–406.

Weitzman, M. L. 1993. What to preserve? An application of diversity theory to crane conservation. *Quarterly Journal of Economics* 108: 157–183.

Westman, W. E. 1990. Managing for biodiversity, unresolved science and policy questions. *BioScience* 40(1): 26–33.

Wiens, J. A. 1992. Ecological flows across landscape boundaries: a conceptual overview. In A. J. Hansen and F. di Castri, eds., *Landscape Boundaries,* pp. 217–235. New York: Springer-Verlag.

Wilkinson, C. F., and H. M. Anderson. 1987. *Land and Resource Planning in the National Forests.* Washington, D.C.: Island Press.

Williamson, R. L. 1973. Coastal Douglas-fir. In *Silvicultural Systems for the Major Forest Types of the United States,* pp. 8–10. U.S. Department of Agriculture Handbook 445. Washington, D.C.: USDA Forest Service.

INDEX

Page numbers followed by "t" and "f" refer to tables and figures, respectively, while page numbers followed by "n." refer to information in notes.